"Here is a philosopher who has learned to think   le body. A keenly aware human animal, Martin Mueller dreams himself salmon flesh. Gill slits open along his neck as he glides between mountain streams and the broad ocean currents. His scales glint and ripple in the moonlight, their reflections posing ever more penetrating questions for our species. This is a game-changing culture-shifting book, ethical and eloquent, opening the way toward a more mature natural science, one that's oriented by our own creaturely participation and rapport with the rest of the biosphere."

— **DAVID ABRAM**, author of *The Spell of the Sensuous* and *Becoming Animal*; creative director, Alliance for Wild Ethics

"The salmon farming industry is not only cruel and environmentally damaging; it threatens to corrode wildness itself. No one has made a more compelling argument to support this fact than Martin Lee Mueller. Philosophically, scientifically, morally, and artistically, Mueller blows the industry guys literally out of the water. If you care about the future of salmon, you must read this essential, rigorously documented book."

— **SY MONTGOMERY**, coauthor of *Tamed and Untamed*; author of *The Soul of an Octopus*

"What if looking at a salmon brought you into deep meditation, and at the end of that meditation you realized that you were looking at yourself, that the salmon was you, you were the salmon, and all is one? That realization is the greatest story on Earth. This book is that crucial meditation."

— **CARL SAFINA**, author of *Song for the Blue Ocean* and *The View from Lazy Point*

"We are slowly realizing—in our dramatic cultural epoch—that dualism has come to an end. Humans do not stand above the Earth; we are but one of its ways of imagining itself. The thinking and feeling of the coming era won't distinguish between imagination, matter, theory, and desire. Martin Lee Mueller's book is one of the first works to radically imagine this new world that is dawning. He shows that reality is a weaving of yearning bodies expressive of innumerable existential stories. Here, outwardness and interiority, humans and salmons, physical descriptions, historiography, and memoir are continuously intertwining. They are equally important aspects of a multifaceted whole that calls for scientific descriptions as well as for personal expressions. Mueller's work is a fine example of the new renaissance slowly gaining momentum, in which we understand our humanness as one strand of the world's manifold desire to become."

— **ANDREAS WEBER**, author of *Matter and Desire* and *The Biology of Wonder*

"This eloquent, impassioned, and often poetic book offers something remarkable: a coherent philosophical and spiritual vision for this era of ecological fragility. Marked by clarity and compassion, *Being Salmon, Being Human* is a beautiful, important work—and a necessary one."

— **JUDITH D. SCHWARTZ**, author of *Cows Save the Planet* and *Water in Plain Sight*

"With this beautiful and important book, Martin Lee Mueller has written a love song to the salmon, and a love song to all life. This book deserves to be read and understood, as an important step in helping us to remember how to love this wonderful planet that is our only home."

— **DERRICK JENSEN**, author of *A Language Older Than Words*,
The Culture of Make Believe, *Endgame*, and many other books

"Mueller's book carries both erudition and urgency secreted within its silvery scales. He understands the hour is late, and his intelligent push towards across-species storytelling is to be taken seriously. Bless his steps, and may his work carry its nutritional goodness far, far over the green teeth of the sea."

— **MARTIN SHAW**, author of *Scatterlings: Getting Claimed in the Age of Amnesia*

"A marvelous exploration of what it means to belong within life's community. Mueller integrates imagination and analysis to produce a book of rare and important insight."

— **DAVID GEORGE HASKELL**, author of *The Songs of Trees* and Pulitzer finalist
The Forest Unseen; professor, The University of the South

"What a fantastic gift from the nation that has given us both deep ecology and farmed fish. Martin Lee Mueller is the first to explain how strange this pairing can be. From Descartes to Naess, he knows his philosophy. But no one before Mueller has dared to ask our *gravlax* itself, 'Who are you?' This is the wildest salmon book ever written."

— **DAVID ROTHENBERG**, author of *Survival of the Beautiful* and *Thousand Mile Song*;
distinguished professor of philosophy, New Jersey Institute of Technology

"How refreshing to read a book on human–fish relations that actually considers the fishes' own perspectives! With lyrical, empathic prose, Mueller beautifully expresses both the sensual world of a salmon and the tragedy of our self-absorption."

— **JONATHAN BALCOMBE**, author of *What a Fish Knows*

"In these pages you will find a well-referenced eco-philosophical story about some of the confounding origins of our separation from both self and all that is nonhuman. Martin Lee Mueller's words are a song of celebration, offering a shared sense of salvation to see salmon and humans, as Haudenosaunee Faithkeeper Oren Lyons might suggest, as relatives rather than resources. Read this book as a clarion call and homecoming for a vision of a new 'Theory of Relatives-ity' with the mantra being: 'Bring the Salmon H.O.M.E. Bring the Humans H.O.M.E. (Here On Mother Earth)!'"

— **BROCK DOLMAN**, director, WATER Institute,
Occidental Arts and Ecology Center

"In *Being Salmon, Being Human*, Martin Lee Mueller brings the abstract categories and arguments of eco-philosophy vividly to life. Weaving together narrative, poetry, science, natural history, and economics, while contrasting Indigenous and modern perspectives on the meaning of salmon, he creates an eloquent, multi-layered terrain of thought and story."

— **FREYA MATHEWS**, professor of environmental philosophy,
Latrobe University, Australia

BEING SALMON BEING HUMAN

Encountering the Wild in Us and Us in the Wild

MARTIN LEE MUELLER

Chelsea Green Publishing
White River Junction, Vermont

The poem "Sleeping in the Forest" on page 90 is from Twelve
Moons by Mary Oliver. Copyright © 1972, 1973, 1974, 1976,
1977, 1978, 1979 by Mary Oliver. Reprinted with the permission
of Little, Brown and Company. All rights reserved.

Excerpt on page 124 is from "The Dry Salvages" from FOUR
QUARTETS by T.S. Eliot. Copyright © 1947 by T.S. Eliot, renewed
1969 by Esme Valerie Eliot. Reprinted by permission of Houghton
Mifflin Harcourt Publishing Company. All rights reserved.

Excerpt on page 185 is from "By Frazier Creek Falls" by Gary Snyder,
from TURTLE ISLAND, copyright © 1974 by Gary Snyder.
Reprinted with permission of New Directions Publishing Corp.

Cover artwork, Salmon Dance Chiin Xyaalaa, by April White.
Please visit www.aprilwhite.com for prints and more information
on April's artwork.

Project Manager: Alexander Bullett
Project Editor: Brianne Goodspeed
Developmental Editor: Rory Bradley
Commissioning Editor: Shaun Chamberlin
Copy Editor and Indexer: Deborah Heimann
Proofreader: Paula Brisco
Designer: Melissa Jacobson

Printed in the United States of America.
First printing September 2017.
10 9 8 7 6 5 4 3 2 1   17 18 19 20 21

Chelsea Green Publishing is committed to preserving
ancient forests and natural resources. We elected to
print this title on 100-percent postconsumer recycled
paper, processed chlorine-free. As a result, for this
printing, we have saved:

48 Trees (40' tall and 6-8" diameter)
21 Million BTUs of Total Energy
4,134 Pounds of Greenhouse Gases
22,421 Gallons of Wastewater
1,501 Pounds of Solid Waste

Chelsea Green Publishing made this paper choice
because we and our printer, Thomson-Shore,
Inc., are members of the Green Press Initiative,
a nonprofit program dedicated to supporting
authors, publishers, and suppliers in their efforts
to reduce their use of fiber obtained from
endangered forests. For more information, visit:
www.greenpressinitiative.org.
Environmental impact estimates were made using the Environmental Defense Paper
Calculator. For more information visit: www.papercalculator.org.

Our Commitment to Green Publishing

Chelsea Green sees publishing as a tool for cultural change and ecological stewardship. We strive to align our book manufacturing practices with our editorial mission and to reduce the impact of our business enterprise in the environment. We
print our books and catalogs on chlorine-free recycled paper, using vegetable-based inks whenever possible. This book may
cost slightly more because it was printed on paper that contains recycled fiber, and we hope you'll agree that it's worth it.
Chelsea Green is a member of the Green Press Initiative (www.greenpressinitiative.org), a nonprofit coalition of publishers, manufacturers, and authors working to protect the world's endangered forests and conserve natural resources. Being
Salmon, Being Human was printed on paper supplied by Thomson-Shore that contains 100% postconsumer recycled fiber.

Library of Congress Cataloging-in-Publication Data
Names: Mueller, Martin Lee, author.
Title: Being salmon, being human : encountering the wild in us and us in the wild / Martin Lee Mueller.
Description: White River Junction, Vermont : Chelsea Green Publishing, 2017.
| Includes bibliographical references and index.
Identifiers: LCCN 2017023754| ISBN 9781603587457 (pbk.) | ISBN 9781603587464 (ebook)
Subjects: LCSH: Other (Philosophy) | Human beings. | Storytelling. | Ecology—Philosophy.
| Philosophy of nature. | Human ecology—Philosophy.
Classification: LCC BD460.O74 M84 2017 | DDC 128—dc23
LC record available at https://lccn.loc.gov/2017023754

Chelsea Green Publishing
85 North Main Street, Suite 120
White River Junction, VT 05001
(802) 295-6300
www.chelseagreen.com

for
Bergljot Børresen
&

Douglas "Runs Like a Deer" McDonnell
(he who lives deep in the forest next to a dark stream)
in gratitude and friendship

You are the story. You won't remember the details, but you know it because you are it.

MIRIAM MACGILLIS

CONTENTS

FOREWORD

*T*his is a rich and timely book. So rich, that I feel overwhelmed in writing this foreword, for it is impossible to encompass its originality, breadth, and brilliance in these few words. My main message to you, the reader, at the outset is that you won't be disappointed by Martin Lee Mueller's phenomenological masterwork into the inner lives of salmon, that most regal of fish who not only feed the flesh of our bodies but also the "flesh of the mind, the landscape of the imagination."

With salmon as his touchstone, as his entry point into our current perceptual and planetary situation, Martin will take you on a series of interlinked journeys: into the heart of the anthropocentric bias in Western culture; into the dream that lead Descartes to propound that schizoid philosophy of his which has so disastrously cleaved us from the rest of nature; into the mind of salmon farmers hungry for profit who care little for the inner lives of their captives; into the very soul of the animistic sensibility of indigenous people that we so desperately need to recover within ourselves if we are to have any chance of surviving the ravages of the changing climate we have invoked; into the lived perceptual field of wild salmon as they travel their turbulent rivers to spawn and die.

These journeys of Martin's are also stories, for this book is a "work about stories we live by, stories within which we come at times to dwell so deeply that we do not easily recognize them as story." Martin's interest is "in exploring how the story of humanity-as-separation is making itself felt in the lived encounter between humans and salmon, and in exploring also how alternatives to that story might already be sprouting in the encounter between us and the salmon." He suggests that the "narrative of separation, of anthropocentrism" pioneered by Descartes's divisive philosophy supresses our own animal nature, and that developing "empathy with other-than-human animals means allowing ourselves to reconnect with those animal-like parts of ourselves" so that we can overcome the Cartesian split and thereby return to our original wholeness.

And Martin certainly knows how to lay down good stories that reconnect us to the living world through salmon persons. His stories and philosophical excursions jump out at you in what I can only describe as a series of powerful, phenomenological "pop-ups"—rather like those 3D books we all loved as children—as his words conjure a three dimensional solidity, a palpable sense of reality about his chosen subjects that leaves me wondering how he could have accomplished such magic with mere words. Here is writing inspired by that other great writer-storyteller-magician, David Abram, whom Martin acknowledges as one of his main inspirations. And following Abram, Martin peppers his text with wonderfully written, highly intelligible tutorials on philosophy, and especially phenomenology, that provide an intellectually nourishing and coherent background to his feelings-based, intuitive excursions into the inner being of salmon.

Another great strength of this work that delights me very much is Martin's care in suffusing his carefully crafted phenomenology of salmon with the latest findings from the sciences of biology, ecology, and even geology so that his accounts of their lived experiences help us to feel what it must be like to be salmon, to navigate an endless ocean, to feel the coolness and taste of the river's water as it flows over our sensory skin, to sense the gravel on the river's base, to chase prey with a quickness of eye and muscle—to suffer as penned salmon do, farmed only for the monetary value of their flesh, the deep innate need for wildness and freedom denied them by their human captors. The science gets really interesting when we learn that nutrients in salmon flesh nourish the soil and trees of the forests surrounding the rivers in which the salmon spawn, spread there by bears and other predators who drag their salmon prey far from the water to feast on the royal fish as they migrate up-river to reproduce.

But this is not just a text about the beauty and wonder of wild salmon. Martin delves into the cruelties and insensitivities of the salmon farming industry to explore how the Cartesian narrative has allowed us to inflict terrible brutality on farmed salmon with no awareness of what we are doing. He tells a narrative in which we are made to live "inside a monocentric story . . . whose baseline is that human rationality—*res cogitans*—is the moral center of the world, around which everything else—*res extensa*—revolves like wandering planets around a great, shining light." I was shocked to read how Rögnvaldur Hannesson, a professor of fishery economics at the Norwegian School for Economics and Business Administration, suggests that it might be necessary to sacrifice all of Norway's wild salmon in favour of their domesticated cousins, purely for monetary gain.

As a counterpoise to the sad and brutal stories of salmon farming in Norway and elsewhere, we hear of the recent removal of hundred-year-old dams on the Elwha River in the Olympic Mountains of America's Washington State, giving the river its wild freedom, allowing salmon whom had almost become extinct to return to spawn and hence to recover their numbers. It was the indigenous Lower Elwha Klallam community living along the river who successfully helped to agitate for the removal of the dams so their sacred salmon could return, restoring both the ecology of the river and also the deep cultural roots and psychological health of these ancient people, for whom the salmon are mythic persons arising out of the Earth's deepest dreaming. Based on Martin's recent visit to Elwha River and its Klallam community, we learn of Klallam lifeways—of their Salmon Boy myth, of their ecologically sensitive salmon fishing practices, of their deep connection to their land, river and salmon. Enlivened by this wisdom, we are gifted with a quickening of ecological awareness within our own blood and sinews.

Martin is a vibrant and important pioneer in a new genre of nature writing that dares to plunge boldly and deeply right into the living, animate heart and soul of nature. His book is sheer delight, a delicious enlightenment. May its message spread far and wide. May it help to wake us up from the mechanistic slumber that now so plagues both ourselves and our lustrous planet.

—STEPHAN HARDING,
Schumacher College, Dartington, UK
July 14, 2017

PREFACE

Ironically, as we work to save the salmon, it may turn out that the salmon save us.

PAUL SCHELL, mayor of Seattle

*W*e inhabitants of industrial civilization still live inside a human-centered story. The story articulates itself in the ways we speak, what we think, how we listen, what we hear. It expresses itself in the physical forms of our lifeworlds, in our legal, political, and economic institutions. It gives structure to the way we conceive of and inhabit both space and time. It shapes our encounters with other-than-human living creatures, as well as with the larger planetary presence. This is the story of the human as a separate self.

The human-centered story is causing the ecological web to come undone at a magnitude of disintegration that is difficult to comprehend, even when one accumulates the evidence. We are in the midst of a systemic ecocide, and the pace of disintegration is so rapid that it is difficult to keep up. The task is no longer to patch up fissures in the story's frayed fabric. This is the time to abandon humanity-as-separation, and to aid forth the emergence of entirely different stories to live by. Even from within this moment of great uncertainty, trauma, and loss, we can anticipate a future in which the human animal can thrive in the midst of the larger, living Earth community. Whatever it takes for our kind to thrive—whatever it takes for this remarkably inquisitive, cunning, creative, and flexible two-legged animal to live a rich and good life—our awareness of who we are as humans must grow from a deeply rooted awareness of the larger planetary presence within which (or whom) we dwell, alongside so many other vibrant presences such as salmon, wolf, moose, alder, elm, mountain, river, or thunder.

This book explores possibilities to abandon the story of humanity-as-separation, particularly as that story has been expressed in the wake of the Cartesian split, which can be thought of as an inaugurating moment of modern philosophy. This involves learning to navigate the story more skillfully, so that we can recognize and observe it, then critique it, and then visualize and realize ways to move through the critique toward alternatives. We must also have a clearer understanding of what is holding the story together, even when the signs of stress are abounding.

The story of humanity-as-separation is thorny and complex, and no single piece of work can hope to tell the entirety of it. Much important work has already been done or is currently taking shape.[1] Far more remains to be accomplished. For still the planet's magnificent web of life is being pierced, fractured, fragmented, poisoned, trashed, consumed, impoverished, killed. Still it is commonplace to speak of the larger living community as an "environment"; still there is a strong habitual bias toward knowing (and treating) other-than-human creatures as "resources" rather than living beings. We are far from having fully envisioned, let alone enacted, a human-Earth relationship that is mutually beneficial. And yet we are moving toward that bold ambition with determination, with inventiveness, and indeed with a sense of wonder.

It was Socrates who called wonder, or *thaumázein* (Θαυμάζω), "the only beginning of philosophy."[2] After him, Aristotle observed that all philosophy arises from wonder, suggesting that it is "through wonder that men now begin and originally began to philosophize."[3] Two thousand years after Aristotle, René Descartes—arguably the first truly modern philosopher—spoke of wonder as "the first of all the passions."[4] Descartes thought of wonder, in the words of contemporary philosopher David Wood, "as a gateway or stimulus to science which could be set aside after serving its purpose."[5] But Wood adds: "My sense is that this is too harsh—wonder need not impede science."[6] And indeed: Even in these days of fast-developing, highly sophisticated, and often highly specialized scientific discovery, we find that each of our encounters with the world has the potential to steep us more deeply in wonder. Four hundred years into empirical science, we might indeed be approaching a threshold in our understanding, where we see the world not as less but as more wonder-full, not as less but as more alive, not as less but as more enigmatic.

This enigmatic encounter with a living planet might also be far more familiar to us than we have supposed throughout these centuries. It might be that our scientific ways of knowing the world are catching up at last with an experience or understanding that has expressed itself in countless forms in various cultures throughout all times: that humans share deep, intimate kinship ties with the larger Earth, and that Earth itself might be more accurately understood not as a giant lump of largely inert matter, but rather, as David Abram has suggested, as our own, larger body. Therein might lie our best hope for crafting an empirically informed, scientifically sound humanism-in-participation. Stepping more fully into Earth's larger and more enigmatic life might be our best bet for becoming more fully human.

What we call wonder is a certain disposition, a certain openness to encountering the world directly through our embodied being, without presuppositions or preconceived theories. That disposition has been most elegantly expressed recently in the school of phenomenology, that study of direct experience. To encounter the world as a phenomenologist is to encounter a world that has not been exhausted of its wonder; it is to encounter a world whose wonder cannot be exhausted, for the sense of wonder emerges spontaneously from within the thick of our direct encounter with the world's never fully fathomable complexity. It can emerge from watching bands of fast-moving rain clouds march above our heads in the evening sky, or from finding the entirety of our awareness absorbed by the sound of splashing water as one salmon after another leaps up a rapid in a turbulent river, or it can emerge from reading a scientific paper that describes in great detail the strangely cyclical, intertwined story of how without life, there would be no more water on Earth, and of how without water, there would be no more life.[7] Or that sense of wonder can emerge from meeting a stranger whose surprising body posture, strange choice of words, eccentric facial expressions, unfamiliar tone of voice, and very different insights give us a glimpse, if ever so briefly, of what it is like to know this same world from an utterly different point of view. Gradually and never fully, the world reveals some of its layered and dynamic complexity, its depth, diversity, creativity, and cycles of participation.[8] It calls upon us to reflect on our place within.

This sense of wonder goes inward in two complementary ways: First, it goes from our sensuous bodies into the depth of the living terrain.[9] And second, it goes into the equally rich, complex, layered, and dynamic topography of our individual mindful bodies, or embodied minds, as we participate fluidly with, and respond resourcefully to, the larger sensuous terrain. Earth might be experiencing itself through us, and Earth might be equally experiencing itself through so many other embodied awarenesses, be they salmon, Sitka spruce, dragonfly, microbe, or whale. There might indeed be a robust scientific case for thinking that now, as Earth has reached the respectable age of 4.543 billion years, this living planet might in some sense be more complexly self-aware than at any earlier moment.

A few words on what this book is, and on what it is not: It is not, first and foremost, a work about the salmon industry (though the industry does figure in it). It is a work about the stories we live by, stories within which we come at times to dwell so deeply that we do not easily recognize them as story. *Being Salmon, Being Human*: The symmetry of the title implies that our sense of who we are as humans is mirrored in our lived relationships with other creatures. It implies also that we become fully human to the degree that we give space to others, such as salmon, to live out their own full potential. My interest is in exploring how the story of humanity-as-separation is making itself felt in the lived encounter between humans and salmon, and in exploring also how alternatives to that story might already be sprouting in the encounter between us and the salmon.

The philosopher Holmes Rolston III has given a luminous description of what ecological philosophy strives for, and how it thinks to get there. He writes that "[ecological philosophy] does not want merely to abstract our universals, if such there are, from all this drama of life, formulating some set of duties applicable across the whole. . . . [It] is not just a theory but a track through the world."[10] This book pursues such a track through the world. It is a way, and if there has to be a "point" to it, then the way itself is it. I seek to wander, as the track unfolds, gradually

out of the story of humanity-as-separation and into the thick of reciprocal participation between ourselves as human animals and the animate
terrain, a thick that we coinhabit not only with salmon but with rocks,
fungi, soil, streams, the pulse of the tides, the oceans. To seek tracks out
of the story of human exceptionalism is not to "leave story" entirely. It is
rather to look for what other stories struggle to be born from the compost of the old, and then to aid these fledglings in their self-emergence,
like a midwife or gardener, rather than an inventor. As Neil Evernden
cautions, "there is no way to deliberately elaborate a new story—it is
not a conscious exercise, not something susceptible of reasoned solution.
One can only hope to pull back and see what emerges to fill the void."[11]

When studying salmon and their awe-inspiring, ancient journey,
one sooner or later recognizes that many journeys have more than one
beginning. So, too, does this book. One beginning was in August 2010,
when the Norwegian business newspaper *Dagens Næringsliv* published
a brief opinion piece on the future of salmon in Norway. The piece was
signed by Rögnvaldur Hannesson, a professor of fishery economics at
the Norwegian School for Economics and Business Administration. In
it, Hannesson publicly raised the question of whether or not it is time
to sacrifice all of Norway's wild salmon in favor of their domesticated
cousins. The paper described Hannesson as one of the country's leading experts on fishery economics, and so his contention held a certain
authority. "We should perhaps ask ourselves what we want wild salmon
for?" writes Hannesson. "If wild salmon get in the way of the fish farming industry, then I must say we must be ready to sacrifice wild salmon.
The industry creates great values and jobs along the entire coast. It is an
important business branch, one that is important to keep. We need not
feel pity for the upper class that will miss a playroom; surely they'll find
some corresponding amusement."[12]

Here I was, living in the land of Arne Næss, founder of deep ecology;
Gro Harlem Brundtland, a pioneer of sustainable development; Fritjof
Nansen, Nobel peace laureate and Arctic explorer; Roald Amundsen,
first man on the South Pole; or Adolf Tidemand, the painter whose
work celebrates Norway's extraordinary richness in natural beauty and
grandeur, its waterfalls, its cascading rivers, its fjords, its high plateaus.

Caring for, loving, and taking pleasure in nature's commons are ideals deeply rooted in the people here; forging kinship ties with these commons might in fact be the quintessence of what it means to make a home in this astonishingly beautiful northern land. And then, from within, this voice that openly and earnestly suggested *letting a species go extinct*! What are the implications of this suggestion? What worldview is being advocated here? What values resonate within it? What would that mean, to give up on the salmon? And how does this relate to the perfect storm of ecological collapse already well under way?

Meanwhile, another beginning was materializing on a different shore, on the Pacific Rim in the American Northwest. There, in a small community along the Elwha River, a short but mighty stream that gushes from the Olympic Mountains and spills into the Salish Sea, an extraordinary story was playing out: Almost exactly a hundred years after the salmon there had been deemed superfluous and in the way of industry, the human community was beginning to help *their* salmon leap back from the razor-sharp edge of extinction and recolonize the river. The return of the Elwha salmon became the motivating force behind the largest dam removal ever to be undertaken, anywhere in the world.

Early in the twentieth century, when white settlers first envisioned the two dams in Elwha River, these dams were to yield "peace, power, and civilization,"[13] according to the pioneer and businessman Thomas Aldwell, who was the chief visionary behind the dams. It was Aldwell's intention to convert the Elwha River "from its waste and loss into a magnificent source of energy and strength."[14] The building of the two dams was widely recognized to be a significant step toward modernizing the American west. In the hands of ingenious humanity, this "mighty power for good"[15] was constructed without fish ladders that would permit the salmon to climb past the Elwha Dam and past the Glines Canyon Dam, and soon the fish, once renowned as the mightiest salmon in the Olympics, began to decline. Meanwhile, the watershed indigenous community, the Lower Elwha Klallam, also found themselves cast into a desperate struggle for survival. Once known throughout the region as the Strong People, they found that the rapid decline of their totem animal, the salmon, went hand in hand with an unraveling of community ties.

But by the year 2013 both dams had been completely dismantled, and the Elwha was running again, free of obstructions. Slowly, the salmon began moving upriver again. And as they did, the human community throughout the watershed engaged in a reevaluation of the cultural narratives that have structured thought and guided action.

In some ways, the events in the Elwha watershed were the antithesis to Hannesson's suggestion. The chronological symmetry between the two cases, combined with their stark contrast, intrigued me. I began reading about the Elwha, and eventually I decided to visit the Olympic Peninsula in the summers of 2012 and 2013, to learn more about the Elwha salmon and about the people who make up the watershed community, both indigenous and others.

But why salmon? The first and simplest answer is I, too, am drawn to salmon! I am drawn to them not only as a thinker, not only as a rational mind studying an "object," but through the fullness of my mindful body. Salmon certainly have provoked my intellectual curiosity, yes. As curiosity turned to professional preoccupation, and as this work stretched out—first into months, then into years—the salmon continued to challenge me, daring me to respond to their call with the best of my intellectual abilities.

But they do so much more! Seeing the salmon come back to spawn in my home river, Akerselva, incites any number of feelings within me: There is a sense of humility as I ponder the incredible journey they have just completed; there is a sense of gratitude for the promise they bring of new beginnings; there is even a sense of ecstasy over seeing such enormous creatures right here in this gushing, gurgling, foaming artery as it cascades right through my city, Oslo.

And of course, I am drawn to the salmon sensually. It takes but the smell of smoked, wild salmon, wafting invisibly under my nose as I sit on a street corner, chatting with a colleague-friend who has brought me this gift from his home river, the Klamath in Northern California, for me to notice how my body instantly responds to the call of salmon. Little flashes of recognition ignite every distant part of my mindful body, setting ablaze my imagination, prompting a sharply focused alertness, making my tongue water, my belly ache in impetuous expectancy. All senses seem to pool into just one desire: to sink my teeth into this flesh, to feel how the finely textured, delicate strip of tail muscle disintegrates in my mouth. To taste that delicious food. To sense how it passes deeper into me. And then, to intuit the metamorphoses the food ignites inside of me: the reciprocal gift of bodies, the dance of belonging, the ancient rite of attraction.

I claim no exclusive ownership to such experiences. On the contrary: Through them I participate in a long and continuous lineage of human ancestors, elders who translated the salmon's epic journey into metaphor; ingenious peoples who carried the salmon's endurance and commitment into story; communities who created fine recipes to celebrate the flesh; sensitive souls who felt the salmon's life cycle resonate deeply with their fascination for the motif of the journey, as well as the motif of transition, and the cycle of birth and decay. For as long as humans and salmon have encountered one another alongside the oceanic rims of the northern hemisphere, their nations have been engaged in a long, enduring, and reciprocal conversation on the good life. For millennia, humans and salmon have struggled to fine-tune their lives in the presence of the other, to flourish side by side, indefinitely. They have sought to creatively adapt to lands in which they were not alone but participants, shareholders, accomplices. Looking back across this vast stretch of time, we dare say that salmon have not only fed the flesh of our bodies, and not only fed the flesh of so many other, more-than-human beings—they have also fed the flesh of the mind, the landscape of the imagination.

For example, in 1653, Izaak Walton wrote that salmon "is accounted the king of freshwater fish," and that he saw in salmon a beauty that "I think was never given to any woman by the artificial paint or patches in which they so much pride themselves in this age."[16] Half a century before, in 1599, the priest and botanist Peder Claussøn Friis wrote that "salmon are deemed the most noble fish and the best and most beautiful that are caught in Norway."[17] More than 2,000 years before Walton and Claussøn Friis, Julius Caesar's legions were marching through the Rhine Valley when suddenly they observed large, silver fish migrating upriver in great numbers—who would not let themselves be stopped even from rapids or minor waterfalls. The Romans, observing the fish in awe from the riverbanks, called them *salar*, Latin for "the leaper." *Salmo salar* remains scientific nomenclature to this day, preserving a link to that original sense of awe or wonder that the Roman legions must have felt when they saw these magnificent fish leap up rapids in one of Europe's mightiest rivers. Go back as far as 23,000 years, and you will encounter cave dwellers in what we call southern France, earlier humans who painted the image of salmon onto a rock ceiling. These ancestors also carved the shapes of the sleek and energetic fish into bones and antlers. As far back as the written, painted, and crafted records of our forebears permit us to follow them into historical time, the salmon were there, gifting our ancestors with their flesh and nourishing our ancestors' imaginations. Our own kind reciprocated the gift by cultivating ways to translate the salmon's nourishment into metaphor, language, and story.

These observations show why salmon are a rather fine topic for a work in ecophilosophy. For if philosophy rises from wonder (as the ancient Greeks were first to suggest), and if salmon have been nudging humans toward experiencing wonder for millennia—prompting our forebears to structure their thought in accordance with these strangely metamorphic fish—then to turn toward salmon in expectation of being drawn once more into wonder is to turn to the cradle of philosophy itself.

I intend to think my way into the topic not as a disembodied, rational intellect pondering an "object of study"—that is, not in the Cartesian tradition—but as an embodied mind pondering the lives of very different embodied minds. Here is a two-legged, lung-breathing, sentient animal at home on the solid ground of the Earth's landmasses, pondering ways to relate respectfully and accurately to scaled, gill-breathing, sentient animals at home in the rivers and the oceans. My body becomes an arena of that confrontation with otherness; I will need to pay attention to the ways in which even the very language I work with is being shaped by the encounter of my mindful body with those enigmatic, intelligent, oceangoing creatures. This entails an engagement with the school of phenomenology, and especially the way it has been articulated by my friend and teacher, David Abram, who was first to bring this branch of philosophy to bear on ecological questions.

Further, I strive to develop this phenomenological work in close rapport with the latest developments in the empirical sciences, drawing on relevant discoveries by ecologists, biologists, and even geomorphologists and geologists. A work in ecophilosophy would remain too speculative if it did not seek a critical conversation with such sciences. Ecophilosophy is distinctly cross-disciplinary, and it might be among those branches of philosophy with the strongest charge to calibrate their reflections, their ways of speaking, their structure, with empirical science. We will see that recent findings and developments in both science and philosophy might now have given us more precise knowledge of salmon than was available to any earlier generation. And we will see that the picture is becoming more wonder-full, not less: the salmon's journey no less awe-inspiring, questions of metamorphosis and interbeing no less potent, the problem of how to live side by side with them no less urgent. Salmon still nourish the human imagination (the scientific, philosophical, and poetic imagination). They still draw the attention toward questions regarding relationship, patterns of organization, economic arrangements, language, and even toward questions regarding the textured experience of time and place. Each of these topics will have its place in the chapters ahead. Throughout, the salmon are going to be subjects in this writing. They are not passive, mute objects; they rather directly prompt aspects regarding the content, structure, and style of this book. This is, at heart, a dialogue between the author of these pages and the salmon themselves.

CHAPTER ONE

Storytelling Animal

*Narrative is the basic modality in which the human
mind functions. We come to understand things
through the sequence of changes that take place and
are best presented in story form.*

THOMAS BERRY

*I*t begins on a frigid November night near the headwaters of the
Danube River in southern Germany. Near-frozen water oozes thick
and black as oil past the pillars of a bridge. Above, the waxing moon
pours glaring white light into the abandoned streets of the town of Ulm.
But below the river's surface, it is almost too dark to see anything. There
is one very large shadow among the others, an absence, a huge blackness
there behind one of the unseen pillars. It does not move. It has not
moved for a long time. It knows when to be patient.

Thirty years ago, when he was just a little fry, he fed on larvae or
insects. But that was then. Now his body is so large that he would starve
on an insect diet. He doesn't know it, but he is a living descendant of
the world's largest salmon. He looks not unlike his closest relatives, the
Atlantic salmon, but he is even larger, and slimmer. Come daylight, his
back will shimmer greenish, his silvery sides will cast a pink sheen, dot-
ted with black spots, and his stomach will be white: the camouflage of a
lifelong river predator. Unlike most of his distant cousins, he has never
migrated out to sea. Neither he nor any of his kin will ever go there. From
the time he fed on the juices of his yolk sac, he has inhabited this river.
It won't be long now until his flesh will feed the insects, the rodents, and
the other fish, perhaps even a fox or a badger. When that time comes, he

still will not have left the veins of this one river, this branching cosmos. As far as he is concerned, this is the way it always has been for the tribe of Huchen, also known as Danube salmon, largest of them all.

He has grown to be exceptional even among his kin. He is now nearly the size of an adult man. When he was younger and his body began to shapeshift, his senses reattuned. There came a time when insects held no more attraction than clouds that passed overhead on a moonless night. Instead, fish of all sizes held the promise of a meal. Trout became a delicacy. A frog might pass as a decent midday bite. Waterfowl were not safe from him. But tonight, no other fish are near. No frogs or newts are around. The fowl are asleep somewhere on the banks. It is quiet under the bridge. Huchen waits and breathes, listens, tastes the water, scans the darkness.

There is a faint plop. A moment's disturbance, a tiny shudder along his lateral line. Huchen knows the sensation: the miniature ripples that brush across his skin, the tiny rhythm beating the surface. Four mouse feet paddle in haste.

By the time mouse has reached mid-river, the large black shadow has calmly positioned itself directly beneath. Then, a splash. Mouse might or might not have time to realize what is happening before she is pulled down into cold, wet shadows.

The splash fades into the winter night and is soon forgotten. No one is out during these frigid hours anyway. The streets of Ulm are deserted. The townspeople are asleep in their bunks. Not far from the river, there is a house, and inside, in a stove-heated room, a young Frenchman is lying in bed. Twenty-three years old (a good bit younger than Huchen), the man is sleeping, and he is dreaming. The year is 1619, and this night will one day be remembered by William Temple, priest of the Church of England, as "the most disastrous moment in the history of Europe."[1] Not that Temple was referring to the mouse—the young Frenchman who happens to be sleeping there in his bed by the stove is René Descartes.

Descartes has three vivid dreams that night. He will later record them meticulously in his journal *Olympica*. It is clear to him even then that those dreams are a visitation, an answer to the problems he has been struggling with lately. The first two dreams are nightmarish. The third is different. There is a book on a table, a dictionary, and then suddenly, there is a second book under his hand, a poetry collection of sorts. He opens it at random, and his eyes are drawn to a phrase: *Quod vitae sectabor iter?* What path in life shall I follow? This dream, as Descartes interprets it, suggests a lifetime's worth of work: nothing less than to reform all knowledge! The dictionary leads him to conclude that all science is one, a revolutionary idea in his time, and he decides that to master this one science would be tantamount to attaining universal wisdom. Descartes reasons that such universal wisdom is something that can only be accessible to the mind of a single man, a man who prepares himself to unify all sciences through a proper method. Once this man has attained and perfected such a method, he can know everything there is to know. This method will lead the way to finding unobscured, unblemished truth.

Eighteen years later, he published his first major work, *Discourse on Method.* The book shows him to be a talented communicator. His was no more perfect, he wrote, than the ordinary mind. His thoughts were not as "quick," his imagination not as "clear and distinct," his memory not as "ample or as actively involved as that of others."[2] He was careful to point out that his *Discourse* was not a textbook, and that he had written it to simply recount the story of how he had conditioned his own reasoning through the method he had discovered. He would merely recount his own history, share the fable that was his.

That fable began long before the fateful night of dreams in Ulm. As a child, young René heard people say that only through literature could one understand anything useful in life. So, young René dedicated himself to becoming a learned man, a man accepted "into the rank of scholars."[3] He attended La Flèche, one of Europe's most famous schools, became a lover of poetry, read the great masters. He studied hard and absorbed the accumulated learning of his age. He acquainted himself with ancient languages, he consulted the fables, he retraced the lines historians had drawn across the map of human memory. He became familiar with mathematics, anatomy, philosophy, law, theology. And yet, young Descartes found himself utterly sobered. Not only did he feel more ignorant than he had before, he also found no evidence that there was anything that was not disputed, anything that was not doubted, anything that might ground the solid and indisputable truth that he was after.

So, the young man rebelled against his teachers. He gave up the study of letters and decided henceforth to consult only the "book of the world." He enlisted as a soldier, traveled, visited courts, and befriended people wherever he went. For several years he was a migrant, a drifter with an urgent sense that he must find something that all his book learning had failed to give him. Yet his years on the road left him just as disillusioned as his years spent consulting the old masters. He realized, to his dismay, that there was as much diversity among the people in the world as there was among philosophers, and that, here as there, he would not be able to find any solid ground.

At last, he resorted to a third and final path of study. He decided "to study myself,"[4] in the hope of finding a basis for all knowledge that would have the certainty and the unity of mathematics. Occupied with these thoughts, he set out for Frankfurt. His goal was to join the ranks of the Catholic army of Maximilian I, the Duke of Bavaria and ally of France, but when he reached Ulm in mid-November, he was held up by the oncoming winter. Descartes spent the day by himself in his stove-heated room, meditating on the disunity and uncertainty of his own knowledge. And then he dreamed, and events unfolded from there.

From the beginning, his theory of truth was also a search for certainty. Central to his search for the right method of attaining true knowledge was a search for the proper *mark of truth*.

How was someone to know whether what he thought he knew was true? According to the empiricist theory Aristotle put forth in his *Metaphysics*, all knowledge is gained and mediated through the senses. This Aristotelian view was widely known in Descartes's day under a slogan Thomas Aquinas had popularized during the thirteenth century AD: *Nihil est in intellectu quod non sit prius in sensu*. Nothing is in the intellect that was not previously in the senses.[5] Descartes disagreed. In his view, human intellect alone could know reality, not through the senses, but through pure and clear introspection into one's own rational thinking. The mark of truth could only be obtained if one withdrew the mind from the senses.

And withdraw he did. In searching for a solid basis of what he could know truthfully and without a trace of doubt, the young Descartes inaugurated a truly radical project, the likes of which no philosopher had attempted for two millennia: He tried doubting nearly everything. He doubted the reality of his sensual experience. He doubted the reality of his own body. He even doubted the existence of the material world. And he found that there was but one phenomenon that he was unable to doubt, no matter how hard he tried: He was unable to doubt that he was doubting! He had peeled off skin after skin of perceived uncertainty—a philosophical exorcism, a disembodiment, radical in its audacity and determination. He was left with what appeared to him the one, pure substance: Thinking. Untainted by a body, liberated entirely and irreversibly from the flesh. That moment of possibility, that enlightenment! That first morning of certainty, after a history of darkness dating back to time immemorial!

Here was his mark of truth. Because he was able to doubt the existence of the material world and unable to doubt his existence as a thinking substance, Descartes concluded that his thinking must belong to a nonmaterial realm that is separate entirely from the material realm. This was the one constant and certain phenomenon. All of this is captured and condensed rather beautifully in his most famous formulation: *Cogito, ergo sum.* I think, therefore I am.

Thomas Berry's Shadow

Descartes's famous phrase is a true milestone in the history of Western thought, helping to forge a particular zeitgeist in thought and language. But that phrase was not communicated in a dry textbook on method; it was presented in the context of a dynamic narrative of personal development. Descartes's success as a communicator lay in his power to narrate, in his effectiveness as a storyteller.

Stories or narratives create the social fabric we inhabit. They teach us how to act, whom we will have compassion for, whom we will ignore, and whom we will confront with aggression. Narratives socialize us into our communities. They inform us of who belongs and who does not. Certain narratives act as baselines, as the foundations upon which our communities rest. They act as the storied strands that always resonate, explicitly or implicitly, in our collective consciousness, in the linguistic habits we craft and nurture, in the technologies we create, in the institutions we conceive, in our actions, even in our inactions.

The human animal is inside narratives. Being in narrative, we are not normally inclined to read narratives *as* narratives. We come to think of a contingent narrative—one possible narrative among several—as the "truth," or as simply "the way things are." But this robs us of the elemental freedom to make an informed choice about whether we wish to live according to that narrative, and whether we wish to comply with its ethical imagination. It also robs us of the possibility to ask whether it is possible to live according to that narrative.

Restoring that freedom and possibility is a task for philosophy. The Pacific Coast philosopher Alan Drengson thinks of narrative language as the foundation of our lived experience. Personal and communal stories, he suggests, create the very conditions of what it means to be a coherent self. Stories frame all our linguistic activities, including both day-to-day exchanges and highly specialized conversations such as astrophysics or mathematics. For this reason, Drengson says, philosophy remains an indispensable intellectual activity, because it offers something different. Philosophy allows us to tell a different story, "the story of our conceptual and cosmic space."[6] This might seem abstract, but it is a temporary abstraction. It will ultimately allow us to inhabit more truthfully the concrete, living world of cascading rivers, calving glaciers, drenching April rainstorms, breathing temperate rain forests, cloud-seeding algae, migrating blue whales, transient wolves, or spawning salmon.

It is toward this sensuous, corporeal reality that my writing gravitates. This book critiques the dominant modern narrative of human exceptionalism, as it appears to be increasingly at odds with Earth's biosphere. Salmon are our lens, our specific path through the world; they help us stay focused, stay on track, even though what is at stake is our very relationship as storytelling animals to this spherical, breathing, and oddly animate planet. This is a particularly vulnerable hour. Earth appears to be slipping into a new geological age, now thought of as the Anthropocene or the "Age of Humans."[7] How can we think clearly, truthfully, and accurately about this hugely potent tremor in the geological history of our planet? How can we anticipate creative, adaptive, and indeed beautiful responses to this rupture in the geological record, one so profound that it will force us to question our most fundamental assumptions about what it means to be human, here inside this biosphere, at this uncertain moment of Earth's very long life history?

As we move ahead, the recognition of humans as "storytelling animals" will be a central concern. The anthropologist James V. Wertsch coined this beautiful and suggestive phrase.[8] Across the span of history, Wertsch suggests, human societies aspire to truth through stories. At any given historical moment, we find ourselves part of numerous stories, of scripts that imbue us with certain characteristics and roles. In Wertsch's view, "there is no way to give us an understanding of any society, including our own, except through the stock of stories which constitute its initial dramatic resources."[9] While all societies, in cultivating their stock of stories, might aspire to truth, there is no guarantee for historical success. Such stories might be told, retold, and refined for centuries. They might gain significant ideological power, including the authority of technology, science, politics, law, and philosophy. They might come to seem common-sensical, reasonable, the peak of a preordained historical development. They might come to seem inevitable, too big to fail. They might even resort to desperate measures to crush anyone or anything that dares to threaten them. But if certain rotten assumptions at their very roots are left untended—or if they remain unnoticed—the noble aspiration to truth cannot, in principle, succeed. Such stories will ultimately fail.

The cultural historian and geologian Thomas Berry has written that the stories structuring our relationship with the biosphere and the larger cosmos are our "primary source of intelligibility and value." Through such stories a person will "come to appreciate the meaning of life or to derive the psychic energy needed to deal effectively with those crisis moments that occur in the life of the individual and in the life of the society." But Berry was perceptive enough to also pay attention to the long shadow cast by such stories. He warned: "The deepest crises experienced by any society are those moments of change when the story becomes inadequate for meeting the survival demands of a present situation."[10] We find ourselves now at just such a crisis moment, a time when our dominant story of human exceptionalism is no longer adequate. But that story, which can be traced back to Descartes, also emerged at a moment of narrative crisis.

Free-Falling out of That Wombish, Local Cosmos

It is not easy to understand how truly militant and original Descartes's skepticism was at the time. Merely to portray him as a rebellious young man who confronted authority would do him injustice. For he appears also to have been superbly sensitive to the general zeitgeist of his age, and bold enough to put his finger into an open, bleeding wound that many felt but few could articulate. Descartes's skeptical philosophy was an attempt to look straight in the eyes of the kind of deep crisis Berry describes. He tried to take a certain existential dread by the horns. But what was that frightening behemoth of his age? How and why did this twenty-three-year-old stand against authority and convention, and declare with such confidence that there would be no more compromises, that from now on the mind would be categorically liberated from the body?

Descartes dreamed his dreams during the second winter of the Thirty Years' War (1618–1648), a war that caused unspeakable suffering throughout Europe. A third of all human lives were lost, and the property damage was massive. Millions fell victim to plagues and famines. Descartes, like his contemporaries Galileo Galilei, Francis Bacon, and Isaac Newton, lived in a time marked by the collapse of medieval certainties, and sensations of vulnerability and insecurity were widespread.[11] The Catholic Church had been the predominant narrator of medieval Europe's story and endowed many of the continent's human communities with meaning. As the Reformation took hold following Martin Luther's protest in 1517, the Catholic Church lost a significant portion of its credibility and began to break up. The breakup was felt as a narrative disorientation. The search was on for a new certainty, a new foundation for knowledge not based on religious faith sustained by dogma. The ambition was to found knowledge on reason alone.

But there was another change of cosmic dimensions also under way. New scientific discoveries were beginning to shape a hitherto unimaginable new sense of reality, contributing to the disruption of previously undisputed, existential certainties.[12] It began some eighty years prior to Descartes's dreams, in 1543. That year, the Polish astronomer and mathematician Nicolaus Copernicus published his magnum opus

De revolutionibus orbium coelestium (*On the Revolutions of the Heavenly Spheres*). Since the days of Aristotle (in the fourth century BC) and Claudius Ptolemy (in the first century AD), Europeans had almost universally accepted a geocentric model of the universe, which placed Earth at the center surrounded by a series of transparent, concentric spheres in which the planets moved. Copernicus's work called that long-standing understanding into question.

The geocentric understanding was itself an elegant response to the cosmic challenge of understanding the movement of planetary bodies above us. The American geophilosopher and anthropologist David Abram suggested to me in conversation that this Earth-centered experience of the universe might have begun when our distant ancestors first learned to walk upright, stretching their curved spines upward freer than ever to look into the night sky. Another of my mentors, the Norwegian writer and veterinarian Bergljot Børresen, thinks that it might have begun even earlier, when our chimpanzee ancestors learned to build tree nests where they could spend the nights safely out of reach of predators. There, they would have had ample opportunity to watch the patterns overhead. Unlike other animals, they learned to sleep on their backs. Looking up into the night sky, they would have likely experienced a sense of astonishment, of wonder!

Wherever the actual origins lie, what seems certain is that once our curious ancestors let themselves be drawn into such wonder in the encounter with the night sky, they would have observed the stars moving in one connected pattern, climbing from the horizon in the east, rising to their zenith, and then sinking again into the horizon in the west. Sooner or later, they would have also realized that not all the lights in the sky were fixed in relation to one another. There were seven celestial bodies that did not seem to fit into the pattern, seven wanderers that seemed to be on their own trajectory. Today we call them "planets," preserving in the word a faint memory of that earlier, primordial wonder that was still very much alive at the time of the early Greek astronomers. The Greek word πλανήτης, or *planētēs*, means just this: wanderer.

No doubt, these erratic trajectories must have presented a serious challenge for formulating a satisfactory theory of the cosmos. The complex model of the geocentric universe was a beautiful way of formalizing the felt experience of standing down here, looking up, and witnessing the entire cosmos revolving on great, celestial wheels. With mathematical exactitude, the Hellenistic astronomers gave formal credence to that one uncontroversial certainty, that sensuous, bodily experience that had been taken for granted since time immemorial: that the Earth underfoot was the center of it all.

Abram suggested to me that this felt experience must have been one of looking up at a wombish cosmos, as if all of reality unfolded inside a great womb. The geocentric understanding of the universe must have been born from an intimate and intense sense of interiority, a sense of living inside something much larger than ourselves. Surely, Abram proposed, the Hellenistic astronomers shared this intimate sense of interiority with many indigenous, oral peoples from other continents. In the Middle East, the night sky was commonly thought of as the canopy of a tent held up by mountains in the four directions. The Pueblo people of North America's Southwest spoke of the sky as a big house, not unlike the five-sided houses they lived in. In the Pacific Northwest, people spoke of the cosmos as a longhouse made of planks of wood. These tales, theories, and traditions might be thought of as sophisticated and ingenious responses of different tribes of storytelling animals to the same, great enigma—the nature of the cosmos. An unbroken thread connected the Greek astronomers and those oral cultures to our ancient chimpanzee ancestors. The collective experience was that Earth was held at the center of a cosmos near and finite, one that wrapped itself in tight, concentric films around that core, holding us safe within. Surely, the geocentric model had been challenged before, by ancient Greek philosophers, Indian philosophers, and Islamic philosophers. But prior to Copernicus, none of the challengers were able to truly contest and challenge the narrative hegemony of the geocentric model.

Then, that wombish cosmos began to convulse and shudder, until it finally erupted. The events that followed were a kind of delivery, a birth, the pains of which we seem not to have recovered from, four hundred years later.

If Copernicus dealt the first blow to the ancient, embodied experience of living inside a wombish cosmos, the next came from the German mathematician and astronomer Johannes Kepler. Kepler formulated empirical laws of planetary motion that gave further credibility to the Copernican heliocentric model. But it was Kepler's Tuscan contemporary, Galileo Galilei, who dealt the deathblow to the old, experiential cosmology. It was thanks to Galileo that the Copernican alternative to the Earth-centered model could be established as a convincing scientific theory. This will eventually bring us back to Descartes.

During a few January nights in 1610, Galileo looked up at Jupiter through a telescope of his own design and observed something no human had seen before: Moons were orbiting around one of the seven celestial wanderers! His observation was robust evidence that not all celestial bodies revolved around Earth. It gave sound, empirical support to Copernicus's earlier predictions and to Kepler's theoretical refinement of the Copernican model.

Despite the strong protests and reprimands of the Catholic Church, which initially succeeded in silencing Galileo, the evidence was now overwhelming. The thought existed—more refined, more empirically sound, and more urgent than ever: Earth was *not* the center of the cosmos. The thought could no longer be unthought. Although Copernicus and Kepler had been mathematical geniuses, able to tear Earth from the perceived center of the universe by the power of their thinking minds, Galileo's genius interlaced mathematical thought with the persuasive power of empirical observation. To Galileo, nature could be described accurately if and only if scientists restricted their queries to quantifiable, measurable data: to numbers, movements, shapes. He rigorously excluded qualities such as taste, smell, color, touch, or even a sense of beauty or compassion from his experiments. His program was a template for reductionist science for centuries to come. Its unprecedented combination of abstract thought with experimentation later prompted Albert Einstein to call Galileo the father of modern science.

Copernicus, Kepler, and Galileo had forwarded a wholly unheard of, wild idea: Earth was moving around the sun, one more wanderer alongside others. This odd and unfamiliar suggestion exploded the earlier conception of the cosmos as a vast inside. The pre-Copernican cosmos had been much smaller, much more finite, much more local. It had been much more familiar to the body. The night sky was a dome arching overhead, and it was not so very far away. Suddenly the cosmos expanded, and it expanded, and to stand face-to-face with that expanding vastness would have been to feel dizzyingly exposed and vulnerable. The sun seems to be rising and setting, the stars and the planets seem to be wandering overhead—had our senses been deceiving us all this time? There was no cover from the vastness, nowhere to hide, nowhere to run, nothing that would stop the vertigo. Descartes's contemporary Blaise Pascal famously said: "The eternal silence of these infinite spaces frightens me."[13] The shadow of a failing narrative that Thomas Berry speaks of truly began to creep across the land. The temperature had dropped; a cold wind had picked up.

Refuge

This is the context for Descartes's existential dread. The European psyche was bleeding from a fatal wound. The overturning of the geocentric model definitively proved that *you could not trust your senses*.[14] Descartes's response to this narrative disorientation was to employ radical skepticism in the search for a new sense of certainty. This is the precise moment, in the thinking of the German phenomenologist Martin Heidegger, when modern metaphysics originated. Heidegger writes: "The metaphysics of the modern age begins and has its essence . . . in the fact that it seeks the unconditionally indubitable, the certain and assured, certainty."[15] Heidegger identifies Descartes as the first truly modern philosopher. He was also the philosopher who definitively split mind from body in the search from that certainty.

With that Cartesian split, a deep ontological crack began to shoot far and fast through the phenomenal world, not unlike when lake ice relieves its inert tension in a rumbling boom that reverberates through the frozen landscape, leaving behind a jagged lake surface. Here, on one side of the split, was the pure substance of his own thinking, that first moment of certainty in the history of thought. Descartes thought of it as *res cogitans*, or thinking stuff: creative, intelligent, self-conscious, self-willed, rational, human mind.

On the other side of the split was *res extensa,* or extended stuff. *Res extensa* included our bodies, the domesticated animal companions with whom we shared our lives, all wild creatures, mountains, rivers, primal forests, the atmosphere, the oceans, the geological forces of the planet, the whole of the planetary presence, the rapidly extending cosmos beyond. The formulation encapsulated and formalized his contemporaries' experience of a cosmos that was suddenly blown wide open. It emphatically acknowledged Galileo's reductionist project, which sought to know the universe truthfully by recognizing all knowledge gained through the senses as deceptive. Never had there been a philosopher who was less ambivalent about the fact that anything other than the thinking human mind was ontologically, epistemologically, and morally irrelevant.

The consequences were profound. Descartes had students vivisect living animals. When the animals on the vivisection table began to scream and writhe and yelp and kick in agony, Descartes urged his students to ignore the screams and to cut deeper into the bleeding flesh. You should not trust your senses: The screams of the animal on the vivisection table might sound to the ears like screams of pain, but the animal is just a machine. All nonhuman animals are machines, apparatuses that can be reduced entirely to their constituent parts, and that can be known entirely through mechanical explanations. Whatever signs of agency, creativity, or conscious decision other animals portray become mere functions of the machine.

Practically in parallel, Galileo and Descartes formalized a withdrawal from the world of the senses in response to the narrative disruptions of their age. To Galileo, this meant that science ought to align itself exclusively with the sober language of mathematics and leave aside the bewildering murmurings of the body. Descartes went even further, proclaiming that the thinking mind would henceforth withdraw categorically and completely from the body, and from all things related to the world of sensory phenomena. The thinking mind would take refuge inside the one thing that had not yet been contested: itself. The clean separation of mind from all else provided a barricade against the encroaching meaninglessness of a vastly expanded space.

The Cartesian split was a desperate attempt to preserve some kind of inside. Sealing the mind off created an impregnable safe haven. This would have been a great relief in those riotous, disorderly times, now that the universe was no longer perceived as a great inside but rather an even greater outside. The fact that the Cartesian split went right through the human body was only consistent. Descartes meant business: If the body had been deceiving the mind all along, it had to be kept at a safe distance. The distrust had to be formalized and institutionalized. The Cartesian split also created a positive vision that charged this new philosophy with a certain mythical power, making it desirable to pursue, an ideal to strive for. With the rational mind, humans could eventually elevate themselves above the muddled realm of the sensuous, up to a plane of no more uncertainty. The fact that humans alone were now left with an interior dimension also implied that humans were not only separate, but superior to the entire rest of Earth's creatures, its landmasses, its oceans. Rising above the confusion was a seductive response. It offered firmness, reassurance, and relief.

Wonder

With the benefit (or agony) of hindsight, it is easy now to condemn Descartes for amputating mind from body so clinically and systematically, and for giving powerful philosophical sanction to animal abuse. We will see how the Cartesian story is still in full bloom today, how it provides an ideological blueprint for the systematic, structural, and industrial-scale abuse of salmon and other animals. We will see the frightening continuity between his philosophical project and modern husbandry practices, and the way it still expresses itself in the structures of our technological lifeworld, in the way we speak about the more-than-human world, and in our political visions.

But this book does not stop with this critique. The intention is to move through the Cartesian story, to look beyond the ongoing objectification of the living creature now manifest in the salmon industry, and to point toward potent, promising, and constructive ways forward in the encounter between us and salmon. For that, it is important that we embrace the Cartesian project fully and complexly from within its experiential horizon, rather than simply cast it off as collateral damage onto the landfill of the history of thought.

What would it have been like to be among the first humans to stand face-to-face with the infinite universe? Would it have been like living a lifetime in a deaf world and then having a boys' choir pierce the eternal silence, unannounced, careless of the effect it would have on everyone's ears? I don't really know. The closest I can come to intuiting my way into the sensation is to meditate on my own birth[16]—the virgin quality of anything I would smell, touch, taste, hear, or see during the first day of my life, the raw wonder at colors as they drift across my vision, or at mother's voice and smell and breath that mingle there before my face, or the sensation of her swollen nipple in my mouth, or the taste of her warm milk as it spills across my tongue, or the warmth and terror and pleasure of father's hand as it presses all of me to his naked chest, warm, dry, sweaty, expanding and then contracting again, deep voice gathering inside and then erupting from the mouth farther up, above me—sensations so intense they make the hair on my just-born arms stand up, so primal and brutal they leave me overwhelmed, free-falling into an unbearable mix of alarm and joy and ecstasy, unable to stop the tears from flowing.

Earth, a wanderer through an infinite universe! What could it mean, being human here atop the crust of this wanderer, in plain view of infinite space? Beside the terror and the fear and the uncertainty, there is the joy, the ecstasy, the novelty, that sense of seeing existence more truly for what it is. We cannot help but acknowledge that Copernicus, Kepler, Galileo, and Descartes were midwives to a true leap in our understanding of the true nature of Being. They boosted a rupture between before and after that was abrupt and irreversible. And of course, they celebrated the empirical method, for it was the final stepping-stone before that audacious leap into a formerly unknown void. The human mind possessed a terrifying and beautiful power, and it expressed its own potential more insistently and urgently at just that moment. Wasn't there something truly awe-inspiring about the realization that the power of reason could transcend the body? It might be at least understandable that the so-called "view from nowhere" became plausible at precisely that historical moment. Doesn't it make sense that Galileo, Descartes, and their contemporaries fell for the belief that the human mind is an all-seeing eye, able to wander freely and without hindrance into infinite space, unbound by tradition, by the body, or indeed by a supernatural God?

This, too, resonates in the formula *cogito, ergo sum*. It is not just the full-blown denial of the body, and not just the full-blown denial of the sentience and intelligence of other animals. It is also a celebration of the empirical method, a salutation to the clarity and depth of understanding that the thinking mind allows. To overcome the Cartesian split, we must endure such complexity and internal conflict.

Huchen

Like many of their oceangoing cousins, the largest of all salmon are facing an uncertain future. At the turn of the twenty-first century, cyanide leaked from a nearby gold mine into Tiza River, a tributary of the Danube. For 250 miles downstream from the spill site, all aquatic life was wiped out. The Hungarian government faced the daunting task of removing eighty-five tons of dead fish from the riverbanks. Other threats, including overfishing, channeling, damming, and pollution, as well as water extraction for industrial and agricultural uses, imperil the magnificent tribe of Huchen. It is not certain whether the Huchen will be able to withstand the many interlocking pressures of the modern life-world and its narrative of human entitlement and separation. There have been attempts to introduce hatchery-bred stocks into several Danube tributaries, but their long-term success is uncertain. And while the fish have been introduced into several waters outside their historical range, all the way down to Spain, Danube salmon have now taken their place on the International Union for Conservation of Nature's Red List of endangered species.

CHAPTER TWO

Hidden Salmon

*Man's freedom lies primarily in the choosing of his
'story,' rather than his actions within that story.*

NEIL EVERNDEN

*A new species would bless me as its creator and
source, many happy and excellent creatures would
owe their being to me.*

MARY SHELLEY, *Frankenstein*

I once attended a seminar at Oslo's newly renovated Mathallen, the city's traditional food market by the banks of the Akerselva. One of the talks was called "The Quest for the Happy Salmon,"[1] given by a technological director of Marine Harvest and a senior researcher at Nofima Marin. As the presentation went on, the title began to sound increasingly cynical. First, the focus was on the industry's growth potential: "Some of the secrets behind the success of the Norwegian salmon industry are these tiny eggs, pearls of the coast," the researcher said. "Almost four hundred million of these pearls are being artificially fertilized every year. They will grow to more than a million tons of fish per year, or twelve million salmon meals every day. Please feel free to be impressed." When it was her co-speaker's turn to speak, he wanted to impress the audience even further. To give a sense of how vast the river of salmon is, he had conceived this catchy metaphor: "Imagine a train," he said. "The train drives past you at a speed of four train cars per hour, twenty-four hours per day, 365 days per year. Each car is filled, to the brim, with salmon. *That* is how many salmon we are producing."

Then the focus shifted to the meaning of a "happy salmon." "Imagine Rica Hell Hotel in Trondheim," he said, projecting an image of the hotel on the screen. The Rica Hell is one of the country's largest conference hotels. "Each of our fantastic pens corresponds roughly to the size and volume of this hotel. These pens are 157 meters in diameter and can have two hundred thousand fish in them, sometimes even more than that. Ten of such pens have a combined value of 300 or 400 million kroners. But what's most important," he raised his voice, "is that the fish have lots of space in them! In fact, only 3 percent of the volume of each pen is salmon, and the rest, all 97 percent of it, is water." He did not say it, but the implication hovered in the room so thick you could have cut it with a knife: With so much space, can the salmon be anything *but* happy?

When I came across these numbers again the day after the talk, I did not know what to make of them. I was in my little office, going through my notes. Pens, the size of one of Norway's largest conference hotels, have 3 percent salmon and 97 percent water. Wouldn't that mean that they are nearly empty? And wouldn't that mean that the salmon who live in them have lots of space to be happy salmon? I have a rather poor numerical imagination, so I sought a way to translate the metaphor. I looked around and thought, well, say this office is a salmon pen. I got up, got a measuring tape, and measured up my office. The room is 2 meters wide, 3 meters deep from door to window, and 2.4 meters high. Hardly more than a cubicle, but it's alright on most days; I have a window and a view of allotment gardens outside.

Next, I thought, if my office is the salmon pen, I would be the salmon. This might work for me. It would help me visualize what 3 percent body mass means in a room that is 97 percent empty. I set to figuring. As it turns out, the formula for calculating body volume is relatively straight-forward. A man who weighs 70 kilograms (which is me, give or take) has a volume of 66.3137 to 70 liters. If this man is 3 percent of a given room's volume, then the entire room would have a volume of 2.21045666667 to 2.3 cubic meters. My office has a volume of 14.4 cubic meters. That meant that to fill my office with 3 percent human and 97 percent air, I would need to invite over *five* other men of roughly my size and weight, plus two children of roughly 10 kilograms each. I was startled, because to imagine that it would take five adults and two toddlers in my cubicle to fill 3 percent of its entire volume was a trying exercise, to say the least.

I had the first result, but I was still not entirely sure what to make of the original metaphor. If it would take five adults and two toddlers in my office for a body-space ratio of 3:97, how many people would I need to bring into the Rica Hell Hotel to get the same ratio? The Rica Hell has 377 rooms altogether, plus several auditoriums, the largest of which can seat well over a thousand guests. That is room for a lot of people. But would the hotel be prepared to accommodate 3 percent humans to 97 percent air? The math was not difficult. If we follow the speaker's suggestion and estimate the volume of the hotel as roughly equal to the volume of a fish farm, with a diameter of 157 meters and a volume of 70,000 cubic meters, and if a 70 kilogram man[2] occupies 3 percent of a room with a volume of 2.3 cubic meters, then the total number of happy people the hotel would have to house is 30,434.

I felt free to be impressed.

Salmon are not necessarily unhappy in such an environment. Perhaps they are happy salmon, or perhaps they are not. It is not an easy conclusion to make. It simply no longer seems clear whether the metaphor is so well chosen, nor whether it says anything about salmon happiness. All it says is that there is a 3:97 ratio between the salmon and their space. To get from there to salmon happiness, the metaphor uses a rather desperate intellectual shortcut. It might be witty rhetorically, but it is not very useful methodologically. I suspect that the way from there to salmon happiness is far less straightforward.

———

Though it is still a relatively recent phenomenon, the global salmon industry is quickly burgeoning into a large-scale international industry. By 2010, Norwegian salmon were being exported to ninety-seven different countries for a collected value of 31.5 billion Norwegian kroner. Just four years later, in 2014, the export revenues had inflated to 46 billion kroner.[3] Salmon has already become Norway's largest "domesticated animal" in terms of biomass. In 2011, the country produced three times more salmon biomass than all other domesticated animals combined. This was just a year after the industry had set a new record: For the first time it had produced, in a single year, more than a million tons of salmon flesh.

Examining the salmon industry will take us into the heart of the human-centered story. For the salmon industry, the fish are known as resource or commodity. There are nods to fish welfare, and to seeing the salmon as living creatures with their own concerns, their need for well-being, their otherness. But we will see that these concessions do not challenge the underlying problem, which is thinking of salmon as commodities in the first place. The salmon's ontological status as functions in a larger industrial complex remains unquestioned. The concessions made to their well-being remain inside a monocentric story. This story's baseline is that human rationality—*res cogitans*—is the moral center of the world, around which everything else—*res extensa*—revolves like wandering planets around a great, shining light.

A Stalled Revolution

Descartes went about his work of isolating the human mind radically from the sensuous world with good intentions. Yet while he aspired, as Martin Heidegger would later, to give a "fundamental critique" of the state of philosophy, it seems that he fell somewhat short. Although he tried to doubt *everything*, Descartes left certain premises unquestioned, even unrecognized—concealed aspects of a larger narrative structure within which the entirety of his work unfolded. In short, Descartes was not radical enough.

The change that Copernicus inaugurated, and that Kepler and Galileo reinforced, was that of decentering the geocentric model of ancient astronomy. It is quite possible that Descartes stopped short of metabolizing the full ontological, epistemological, and ethical implications of this decentering. Although an outrageously different idea was taking shape, Descartes carried a certain residue of the old monocentric story into his philosophy. Even in a vastly decentered universe, he insisted that there remained a single, central access to truth—the human mind. The Heidegger scholar Charles Guignon has fittingly written: "Starting with Descartes, a clear distinction is drawn between what is given in the mind in perceiving, willing, imagining, desiring, and other mental acts, and what exists in the external world and is represented by such mental acts. *The subject becomes the center around which all other entities revolve as 'objects' of experience.*[4] Mind becomes the new gravitational center: All else "revolves" around mind as "objects of experience." What if the Copernican Revolution still hasn't been completed? What if it will only be completed when we have successfully overhauled the old ideology that humans are at the center of Being? And what would it take to bring the Copernican Revolution to its completion? What would it take to fully decenter the narratives of who we are as storytelling animals here inside this Earth, and to recognize and appreciate the many other storylines whose strands are woven into the delicate web of Earth's biosphere, the short-lived and the ancient, the feeble and the grand?

Descartes's skepticism and the Copernican Revolution ushered in a new era in human history, commonly designated the "modern" age. But what does that mean, "modernity"? The Australian philosopher Freya Mathews writes that "the hallmark of modernity is radical change—in the form of development, control, management, design, intervention, progress, improvement, even salvation."[5] Derived from *à la mode* ("of the present"), modernity fetishizes the need to keep up with the latest: "Modernity is that period which can be characterized in terms of its commitment to the ever-emerging new, its dissatisfaction with the given, its radical discontinuity with the past and its dissociation from tradition."[6] In the very name given to the historical period that began during Descartes's era echoes the aspiration to reveal the world in a mode of radical discontinuity and change, as a succession of the ever-new. But Mathews's etymological reading of "modernity" misses a certain inconsistency in that aspiration. Modernity *conserves* the monocentric bias of pre-Copernican days, in that it defines mind as that around which all else revolves. In that, modernity is both radical and conservative.

Take the mechanistic metaphor, the world-as-machine: The American writers Charlene Spretnak and Carolyn Merchant both observe that no narrative tradition before modernity had ever considered the world to be a complicated machine. This reinterpretation of the world from a living being into a machine was unprecedented and unreservedly original.[7]

But there is also that conservative aspect of modernity, that monocentric residue. The political theorist Michael S. Northcott from the University of Edinburgh writes: "The modern world is often described as humanist, but its announcement by scientists and philosophers required 'the simultaneous birth' of 'non-humanity'—things, or objects or beasts. . . . The modern constitution, precisely because it is rational artifice, requires the separation of nature and society."[8] In a truly modern age, nature can never be anything but backdrop, *res extensa*, environment—meaning literally, "that which surrounds," "that which orbits around the one center."

Modernity might never have been as modern as it thought itself to be.[9] At best, it has been selectively modern, overhauling only that which was inconvenient, yet retaining that which ensured it a continued sense of entitlement and power. The predominant ethical stance of modernity remains anthropocentrism, which stands directly in the tradition of Descartes's monocentric thought. Modernity's humanism is humanity-as-separation. The premise of human exceptionalism has never been sufficiently renounced in that tradition.

This observation would be less remarkable if modernity did not still entertain a self-image of foundational discontinuity. This discontinuity is present in the idealization of the storyline native to modernity: progress. Modernity reflects the ideal of liberation from traditional and natural limits, and a shift in focus to the ever-new. In this lies a certain epistemological optimism, the idea that one "right" story would emerge out of the murky haze of medieval "worldviews," in a process of technological and moral "progress."

———

What is a happy salmon? In the presentation I saw, it was suggested that "happy salmon are those who are contained for life in near-empty, huge pens," but details about the reality of the salmon industry make this implication untenable. The veterinarian Siri Martinsen, director of the Norwegian animal rights organization NOAH, writes: "Farmed fish live lives far removed from their instincts and needs. . . . Abnormal conditions and high densities can lead to 'wear and tear' on the fins and nose as a result of aggressive encounters between fish. Congenital defects and serious injuries as a result of disease are not uncommon. Increased mortality, viruses and bacteria are part of the fishes' everyday life."[10]

The journalist and author Niels Christian Geelmuyden has also written a bleak description of the salmon industry that deserves to be cited at length:

> Many juvenile salmon are effectively blinded upon contact with salt water because their eyes have not had time to adapt. Research estimates that 80 percent of the salmon in Norwegian plants suffer from cataracts, and that half of these are fully blind. What makes the visibility—and likely the smell—so bad are the feces from the 199,999 co-inmates in each pen. Not to mention the millions of salmon who reside in the nearby pens. The total equivalent of all of Norway's fish farm feces is the annual sewage of 11.9 million people, or, in other words, roughly the entire population of Tokyo. . . . A growing number of farmed salmon is now struggling with deformities—a result of inbreeding, of damages incurred through vaccination, and of artificial seasons, which accelerate the fish's overall maturation but do not give their bones a chance to keep up. The salmon incur permanently bent-down lower jaws, pathologically shortened upper jaws, hunchbacks, exposed gills, and truncated tails. More prevalent than these, but also more severe, are case of deformed internal organs.[11]

The deep optimism inherent to the story of humanity-as-separation leaves cripples and corpses in its wake. If the story's premise is the categorical divide between rational, thinking minds and everything else, there will always be losers. Salmon are slaughtered in especially designed slaughterhouses. During transport to the slaughterhouses, they can be cramped together in densities of about 150 kilograms of salmon per cubic meter of water.[12] To achieve the same ratio of bodies to space, it would take 150,000 people (more than the population of the city of Stavanger), packed into the Rica Hell Hotel!

Then again, perhaps the way from body-space ratio to happiness is every bit as straightforward as the speakers implied. It merely depends on what you mean by happy salmon. For the speakers, happiness was connected with growth rates: "We dare say that those animals that produce the largest weight are also the ones that thrive best. Much of what we are occupied with is to choose those animals that thrive in confinement." The faster an animal grows in captivity, they say, the clearer it proves to be a happy animal. The happier it is, the more profit we make. Incidentally, this circular logic also means that salmon happiness can be directly equated with biomass. Measured by that standard, contained salmon are getting happier every year.

Some Tentative Steps Back Inside

While these realities continue to shape our lifeworld, the philosophical community has tried to address the root causes of these modern practices. Martin Heidegger published his groundbreaking book *Being and Time* in 1927, three and a half decades before the ecological debate would emerge fully into public consciousness.[13] A conversation with *Being and Time* can fuel our phenomenology of story in many ways. Here, Heidegger anticipates a key argument that later emerges in ecophilosophy: that the ecological crisis—this moment of comprehensive, sustained, and escalating ecocide—signifies a crisis in our relation to the world as human animals and moral actors.[14] Heidegger's unique angle was to suggest that the history of philosophy can be read as one of accumulating layers, or strata, of metaphysical interpretation. These philosophical strata each offer their own view of what it means to be human here on Earth, in relation to this atmosphere and these forests and this soil. According to Heidegger, accumulating layers is no guarantee of an increase in wisdom. In fact, it can have the opposite effect: Successive interpretations might overlay each other and gradually shroud or distort a more original experience of Being, and of ourselves in relation to the world.

His response was to develop a "fundamental ontology," a philosophy that would deconstruct or remove this accumulated metaphysical debris. It was a peeling of the onion, if you will, and it would gradually loosen the grip of those received philosophical interpretations and the way they distort our spontaneous and immediate experience of Being. Descartes's withdrawal into the thinking mind might be the hardiest and thickest layer to remove. Only by removing it, Heidegger suggested, can we penetrate to the "darkest" term, Being itself.

Heidegger began his ontological tour de force with an epistemological localization: that we are already in the world whenever we venture to speak about that world. Being human should be thought of as being-in-the-world! The phrase might feel bulky and awkward to the tongue, but it is an excellent move. Heidegger breaks with Descartes's withdrawal from the tactile, sensuous world. He questions the idea that the human mind is the only true inside left. If we are already in the world when we speak about it, then "the world" must be an inside. Being human, we are "thrown into" the world. And being thrown, being entirely within and entirely thereof, our primary access to clarity and truth would not be to step outside. It would rather be to ask how precisely it is that we are in this world. What are the qualities of that interiority? What are its textures, its patterns, its rhythms? With Heidegger, this world becomes once again the context and horizon of our existence, and being human must mean being in relation to this Earth. By designating being human as being-in-the-world, Heidegger paved the way for radically rethinking the question of being human: from within this atmosphere, within this windswept, rain-soaked planet as it journeys around its star.

Say that I am impatient and want to know without further ado: What—who—is a salmon (let alone a happy salmon)? Who is that weirdly metamorphic fish in the rivers and in the oceans, or even in the plastic holding pens out in western Norway's fjords or the inlets and bays of British Columbia? How do I encounter it most truthfully? Heidegger's teacher, the German philosopher Edmund Husserl, used the phrase "to the things themselves!" (*zu den sachen selbst!*). In Husserl's view, encountering salmon would be straightforward and unambiguous.

But Heidegger sounded a warning against his teacher. He insisted that phenomena do not necessarily present themselves unambiguously and directly. There are metaphysical layers that might distort our experience of salmon so profoundly that salmon herself—the living, gill-breathing, leaping fish with her unique experience as a sentient creature, with her own rich and complex sensations of pain, hunger, sexual arousal, excitement, curiosity, or joy—might be crushed under the weight of these layers. We might have been socialized so deeply into a certain narrative that we are unable to perceive salmon as anything but machinelike, anything but resource, biomass, object for economic speculation, or vacuum-wrapped food portions in the grocery store. We might even be directly in the presence of living salmon and never recognize the life of the creature! Guignon writes of this: "The objects that show themselves at the outset are . . . not the genuine phenomena with which phenomenology deals. The 'phenomena' of phenomenology are precisely what do not show themselves."[15] The phenomena withdraw into concealment and linger there—in plain view and yet enigmatically out of sight.

When Stories at Once Reveal and Conceal the World

The dominant story of humanity-as-separation has both revealing and concealing aspects. In revealing the world to us in a certain way, the story encourages certain ways of speaking, certain ways of building and using technology, and certain ways of engaging with the more-than-human world. Think of the animals on Descartes's vivisection table. Nonhuman animals are on the far end of the ontological divide; they are "outside"; they are "other," which means they cannot be said to possess any capacity for self-conscious experience; all that is "outside" functions according to mechanistic laws; animals are machines; their screams do not indicate to us any moral dilemma. If humans alone are moral agents, then others cannot participate in the moral circle as moral subjects with intrinsic value. At best, they can be treated as moral objects with instrumental value.

The machine metaphor implied that you should not trust your senses. On the basis of this distrust, philosophical authority helped institutionalize a certain moral degeneration: The senses might be provoking a feeling of empathy with the screaming animal, but that is not relevant, not true. Your experience of empathy with the suffering animal is giving you an incorrect perception of reality. If you wish to be a respected member of the scientific community, you must actively unlearn your spontaneous experience of empathy, with all the consequences for your own actions, as well as for the lives of those animals.

In Heidegger's later work, he argues that technology in the modern metaphysical tradition tends to "provoke," "force out," and "challenge forth" the world, such that the primary intention is to "force" the world to yield energy. Heidegger calls this "the essence of technology," and it is this essence, he says, that constitutes the "culmination of modern metaphysics."[16] The philosopher Bruce Foltz describes how this perceptual rupture plays out in Heidegger's work: "'[T]o be' is 'to be a resource,' that is, to be 'in stock,' in supply, ready for delivery."[17] To say that modern metaphysics culminates in technology is to say that there are tight and dense entanglements between the modern narrative of separation and technology. The perfectly straight concrete road cut through a boreal forest is a thought structure *concretely* articulated, and in turn it seeds the imagination with opportunities of how to encounter (or use) that forest in ways unthinkable—and thus unreal—without the road. Likewise, a dam built into a salmon river is abstract thought physically articulated; in turn, the presence of the physical structure gathers the imagination around particularly mechanistic, utilitarian opportunities for encountering the river. Narrative and technology reinforce one another in powerful feedback loops, each contributing their share to revealing the world in a particular manner.

The complementary concealing aspect of narrative is now not so difficult to grasp. As Guignon describes it: "When a worldview becomes firmly entrenched, it tends to perpetuate a set of problems that are taken as natural and obvious. The possibilities of thought become calcified; the same questions and the same types of futile answer are repeated along the guidelines laid out by the grid that structures our thought."[18] As a worldview, human exceptionalism scripts certain ways of being in the world, making these ways appear ordinary and unproblematic. In the age of modern metaphysics, fish, whales, sea turtles, rivers, or even the very oceans are all revealed as a stockpile to the rational mind. In such a world, moderns come to believe, in Heidegger's words, that "we are such terrific people, the Lord must have given it to us in our

sleep."[19] Being in narrative, we cannot easily read human exceptionalism as just that—a story that has certain historical foundations and
belongs to a certain historical and geographical context. We might not
be determined to remain inside the scripted horizon of those narrative
structures, but it is not at all easy to claim the larger freedom available to
us, namely the freedom to question the deepest premises by which we act
and by which we understand the world.

In the case of human exceptionalism, the story conceals the possibility
that there might be alternatives to encountering the more-than-human
world through the lens of mechanistic thought. It also conceals the
possibility that there might be (self)consciousness, intelligence, and
agency beyond the species divide. Its sheer ubiquity might also conceal
other narratives, different stories that can function as primary sources
of intelligibility and value. This story's stated ambition is to cast the
entire world into its metaphysics. It aspires to reveal the entirety of the
larger planetary presence as what Heidegger calls a "standing reserve"
(*Bestand*). Other-than-human creatures, as well as other-than-modern
cosmologies, become concealed in their uniqueness, and in their ability
to draw the imagination out of the gravitational force field of that one
story.[20] If the intention here is to empower ourselves to navigate cultural
narratives with a mature power of judgment, then it becomes important
to gain a clearer understanding of what there is to choose between. It is
a fundamental concern. As Neil Evernden has pointed out, our freedom
lies not primordially in our actions within a story, but rather in the
choosing of our story.

Ontological Schizophrenia

In the spring of 1969, a small Norwegian company called Mowi set their
first salmon fry in a pen off the coast of the island of Sotra, outside
the city of Bergen. In the half century since, the industry has gone
from being the backyard-enterprise of a few daring pioneers to being a
multi-billion-dollar industry, producing three times as much food as all
Norwegian livestock producers combined. Growing the grain products
that are processed into salmon feed requires the equivalent of all of

Norway's agricultural areas. While it remains unclear where the industry is headed, the changes now under way are rapid and unprecedented. The trend is to increasingly accommodate the living creature to market logic, channeled by technological capability, and validated by an anthropocentric realism. "Ideally," writes anthropologist Anne Magnussøn in her 2011 dissertation on salmon aquaculture, "there is a wish for absolute control with the fry's built environment."[21] The ideal of total control is by now being extended so far that it has taken over the salmon's genetic memory. A small company called AquaBounty Technologies of Waltham, Massachusetts, recently made history when their so-called AquAdvantage salmon became the first genetically modified (GM) animal intended for food ever to be approved by the US Food and Drug Administration.

AquaBounty Technologies has found a way to make GM salmon grow to market size in half the time it takes competitors. They have invested twenty-plus years of research and $60 million to develop the necessary technology. The technology is complicated, and practically speaking, it is also rather successful: First, you take the antifreeze promoter from a creature called ocean pout. Ocean pouts are eel-like fish who live in the Northwest Atlantic, off the coast of New England and Canada. Antifreeze proteins in their blood enable the fish to live in near-freezing waters. The proteins are made by a promoter that is basically an on-off switch, and because ocean pout live where they live, the switch is always "on" to ensure a constant supply of antifreeze hormones. Having extracted this on-off switch from the ocean pout, you splice it with a Chinook salmon growth hormone gene. The result is a Chinook growth hormone gene that is perpetually switched on and that will never cease generating growth hormones. Finally, you take this new gene and transplant it into Atlantic salmon. You end up with a creature that outwardly resembles Atlantic salmon, but whose inward metabolism is a hyperactive, restless composition that strives unceasingly toward one end—rapid growth. The creature is so novel that AquaBounty decided it was best to patent it.

AquaBounty submitted an application to the FDA to approve this new product, their GM salmon, to the US market in the fall of 2010. Their application went hand in hand with a meticulously orchestrated media campaign. Throughout the many interviews, articles, and video clips that appeared both in national and international publications during that time, one repeated sentence stood out. Take as an example the following statement by Ron Stotish, the chief executive of AquaBounty Technologies: "In every measurement and every respect, these fish are identical to Atlantic salmon."[22] It is a peculiar statement. At the very least, it is simply counterintuitive. If you take a salmon and implant a distant cousin's growth hormone into it, how can it be the same as it was before? It also stands in direct opposition to another statement, also made by AquaBounty in the fall of 2010, that the GM salmon *cannot* be assumed to be identical" to the non-GM salmon, "as some differences do occur."[23] These fish are both defined as identical in every measurement and every respect, *and* they cannot be assumed to be identical!

What to make of this? The apparent contradiction is actually a careful strategy. During the early 1990s, the first genetically modified plants were being developed, and companies with an investment in this emerging market expected a return on their investment. On July 17, 1991, Michael Taylor was appointed as the Food and Drug Administration's first deputy commissioner for policy. The position had not existed before. His job was to oversee legal concerns regarding existing genetically modified organisms (GMOs), and to develop legal guidelines for future organisms with altered genetic memories that would be introduced to the market. Taylor had worked for the FDA before, between 1976 and 1981, when he had served as the executive assistant to the FDA's commissioner. Then, in 1981, he went into private law practice, where he founded and led his law firm's "food and drug law" practice. One of his most important customers was Monsanto, and Taylor played a key role in legalizing Monsanto's GM bovine growth hormone (which has since been banned from use in Canada, Australia, New Zealand, Japan, and Israel, as well as by all members of the European Union). During his tenure as deputy commissioner of policy that began in 1991, the FDA ratified the so-called "principle of substantial equivalence." It says that any new food—genetically modified or not—must be considered the same and as safe as conventional food if it shows the same composition and characteristics as the ordinary food.

Some have defended the principle of substantial equivalence as a well-designed instrument for deciding which foods are safe and which are not. In an article titled "Substantial Equivalence Is a Useful Tool," published in *Nature*, Peter Kearns and Paul Mayers argue in the principle's favor.[24] Others have criticized the tool for being so lax as to be essentially useless, because there is no binding agreement as to what precisely makes one organism substantially different from another. Erik Millstone and his colleagues write: "The concept of substantial equivalence has never been properly defined; the degree of difference between a natural food and its GM alternative before its 'substance' ceases to be acceptably 'equivalent' is not defined anywhere, nor has an exact definition been agreed by legislators. It is exactly this vagueness which makes the concept useful to industry but unacceptable to the consumer."[25] The Royal Society of Canada expresses a similar concern: "Substantial equivalence does not function as a scientific basis for the application of a safety standard, but rather as a decision procedure for facilitating the passage of new products, GE [genetically engineered] and non-GE, through the regulatory process."[26] In other words, the Royal Society considers the principle's foremost purpose to be moving new products quickly toward approval without causing much regulatory stir. It allows for new products to be approved within months of being developed, whereas comparable products in other industrial sectors require stringent, expensive tests conducted over the course of years before they are approved or rejected. By implication, FDA policy since the beginning of the 1990s can be read as a synergetic and progressive deregulation of the GMO market.[27]

AquaBounty's decision to brand their GM salmon as "identical in every measurement and every respect" to other Atlantic salmon now becomes clearer. AquaBounty knew that legal as well as commercial success is most strongly implied by the principle of substantial equivalence. They knew also that no case brought to the FDA had ever been declined on account of substantial *difference*.[28] Their Janus-faced public relations strategy appears to be a simultaneous appeal to FDA guidelines and to public concern. The precise wording might change, but the message does not: "The chinook growth hormone is the same as the Atlantic salmon growth hormone; it is simply regulated differently. Their ability to grow faster does not change the biological makeup of the fish."[29] Or: "Triploid eyed-eggs for *AquAdvantage* Salmon are produced in a manner that results in the culture of an all-female population of triploid fish that are otherwise substantially equivalent to farmed Atlantic salmon."[30] Or: "AquAdvantage® Salmon (AAS) include a gene from the Chinook salmon, which provides the fish with the potential to grow to market size in half the time of conventional salmon. In all other respects, AAS are identical to other Atlantic salmon."[31] But the constant repetition of the same paradox does not make that paradox any less problematic. Genetically modified salmon are genetically modified salmon. Having been modified, they are not identical with other Atlantic salmon. Vandana Shiva, the Right Livelihood Award laureate, has come up with a name for this type of paradoxical assertion: "[when they want to patent these products], they say these are 'novel,' these are not natural. But when it comes to safety, they say: it's just like nature, exactly as nature made it. I sometimes call this 'ontological schizophrenia.'"[32]

And yet, and still, I wish to know: What or who is a salmon? It seems that we have only just begun to sound out the question in its depths.

CHAPTER THREE

Exploited Captives

*Though we have life, it is beyond us. We do not
know how we have it, or why. We do not know
what is going to happen to it, or to us. It is not pre-
dictable; though we can destroy it, we cannot make
it. It cannot, except by reduction and the grave risk
of damage, be controlled. It is, as Blake said, holy.
To think otherwise is to enslave life, and to make,
not humanity, but a few humans its predictably
inept masters.*

WENDELL BERRY

*T*he anthropologists Marianne Lien of the University of Oslo
and John Law of the Open University once visited an industrial
salmon rearing unit where five hundred thousand fish were being raised
to market size. For comparison, five hundred thousand fish is roughly
equivalent to all free-roaming Atlantic salmon who returned to Norway's
coast in 2010, the year the authors wrote their piece. When they visited,
the fish were approaching their slaughter weight of about 5 kilograms
each, making the entire biomass in the pens 2.5 million kilograms.
Numbers make up the beginning of Lien and Law's discussion of the
fish farm industry. This is not a coincidence. Biomass is how fish farms
define what a salmon is. The metaphor of *biomass* is a stand-in for a
concrete equation: Living beings equal flesh; flesh equals mass; mass
equals numbers; numbers equal economic performance.

From the perspective of head offices, the crucial thing to consider is
feed. Feed constitutes nearly two-thirds of the running costs of industrial

salmon production, and salmon are specifically bred to be hungry.[1] A central task at any salmon production site is to facilitate a ceaseless conversion of feed into flesh. The more efficient the conversion, the better the business. This is measured in the industry by the so-called feed conversion ratio, or FCR. To achieve the best FCRs, farmers across the country are streamlining their on-site operations with high-tech infrastructure.[2] The humans on-site monitor the machines, run statistical data through the machines, maintain the machines. Machines become the matrix by which salmon are perceived, conceptualized, handled. In the context of a modern fish farm, salmon are defined by numbers, numbers are gathered together into statistics, and statistics make the fish, the living creature, into a variable in a mathematical equation. Salmon become inherently knowable, manageable, controllable.

Lien and Law ask, "What is a salmon?" Their question is meant to motivate a survey of the various ways in which humans interact with salmon. It lets them "attend to how salmon are *done* in practice."[3] They assume that scientific classifications do more than depict the world objectively and truthfully; such classifications also help shape and condition the very circumstances they describe. Scientific classifications problematically tend to establish a "*universalising* discourse" and to "enact universal knowledge."[4] Salmon, from this view from nowhere, become an unambiguous category, defined by feeding patterns and numerous phenotypical, morphological, and genetic markers. At the bottom of it all lies a solid epistemological certainty concerning what, or rather who, salmon is.

But Lien and Law suggest that it is not that simple. Salmon are not "given in the order of things."[5] Salmon—the fish, the living beings, the creatures we can enter into relationship with—are not a clear and unambiguous category; they come into being differently according to who interacts with them. Fish farms marshal salmon into one overarching category: biomass. From an anthropological point of view, "what salmon are" depends on how humans interact with them. The question, "What is a salmon?" fans out into a more heterogeneous, complex, and at times incoherent "patchwork."[6] Your answers to this question will be far from universally valid, and far from a generalized view of "the" salmon. What we consider knowledge is far less certain, and far more embedded, than we might originally think. As certainty is shaken, and as generalization gives way to embeddedness, apparently clear-cut and sharp boundaries might turn out to be permeable horizons.

The dichotomy "wild versus domesticated salmon" is one such horizon, according to Lien and Law. Our languages did not begin making the distinction between wild and domesticated salmon until the emergence of the still-adolescent salmon industry. But Norwegians were getting actively involved in the fate of salmon in the 1860s, a hundred years earlier. Back then, people were experimenting with breeding fry in riverside hatcheries, taking broodstocks from some rivers to release fry into others, and releasing millions of the tiny fish into their waterways. When people have been dabbling in salmon reproduction for one hundred and fifty years, not just forty or fifty, it becomes difficult to draw a clear line between a wild salmon and a domesticated one.

The main problem with the distinction might lie in the way domestication is being defined. Juliet Clutton-Brock of the British Museum's Department of Zoology defines a domesticated animal as one who is "bred in captivity for purposes of economic profit to a human community that maintains *complete mastery* over its breeding, organization of territory, and food supply."[7] But the salmon industry, according to Lien and Law, involves so much uncertainty and unpredictability that it is problematic to speak of complete mastery. Salmon inside cages have remained "elusive creatures," in their words. Salmon who break out of their cages become even more elusive, and once escapees manage to swim to a free-roaming population's spawning ground, do their dance together with wild salmon, and lay their eggs in the gravel, the notion of complete mastery becomes insupportable. Feedlot salmon remain essentially "slippery," making the boundaries between them and wild salmon unstable.[8]

Clutton-Brock's definition of domestication also implies that domesticated animals have no agency over their own lives, no possibility to act on the basis of their preferences, distastes, or compulsions. But feedlot salmon, according to Lien and Law, do have some agency.

To prove their point, they describe how a feedlot's caretaker will typically make several hand-feeding rounds every day. The caretaker stands by the cages, casts handfuls of pellets into them, and observes the salmon's behavior to see how eagerly they are feeding. It takes skill, and an untrained eye would have trouble making sense of the frenzy of slapping shadows and splashing water. Lien and Law say it is important to keep in mind that these hand-feeding rounds contribute little to the overall conversion from feed to flesh. And yet they do serve a purpose: They act as "an invitation for the fish to come close and make themselves known." The caretakers observe the fish and judge whether they are still hungry or not. The agency that Lien and Law locate here is the salmon's ability to accept the food or reject it: "[S]almon is done not simply as a hungry animal," they write, "but as an animal willing and able to satisfy its own hunger. In other words, the animal enacted through these practices is not a passive entity, but an animal with a certain form of agency."[9]

Having challenged the implications of Clutton-Brock's definition of domestication, Lien and Law suggest an alternative definition, one they consider truer to the fluid nature of salmon domestication. "Domestication," they propose, "is an ongoing and unruly relationship."[10]

Domestication as an ongoing and unruly relationship: This definition keeps us wary of circumstance. But it also thins out the concept of domestication to the point where its effectiveness as an analytical tool becomes doubtful. What is the analytical gain of a definition that pushes the doors open so wide that nearly anyone can wander in or out? The problem with Lien and Law's alternative to Clutton-Brock's definition is that it takes the relativizing too far, to the point where the definition can hardly make any meaningful statement about why it is useful to distinguish between domesticated and undomesticated salmon.

The modern English forms *domestic*, *domesticate*, and *domestication* all came to us from the Latin *domus*, or "house." They all have to do with houses, and with people, for people make and inhabit houses. Domestication also involves a certain kind of permanence in the relationship between people and other animals. Houses are commonly made to be permanent dwellings to which people and their domesticated livestock can return.[11] This is simple enough.

But domestication is also nestled firmly within a semantic minefield. The nearness of the word *domestic* to *houses, dwelling,* and *home,* for example, complicates our understanding. Think of the ways that the Greek word for house, *oikos,* lives on in modern use. Both *ecology* and *economy* are derived from it, and the semantic neighborhoods of these two words are so foreign to one another that it is not easy to think of, let alone reconcile, their common heritage. "To domesticate" can be used to describe how something or someone is being "caused to be at home, naturalized."[12] If domestication describes the act of taking creatures "out" of a natural state and "into" a realm defined by humans, then connoting *domesticate* with the verb *naturalize* confusingly both upholds and undermines the old dualism between humans and nature. "To domesticate" can also mean "to attach to home and its duties." Duties to whom? If duty is defined as "the action and conduct due to a superior," then it would seem that domestication does indeed point to an uneven power relation. One party (presumably humans) is understood as superior to the other party (presumably the salmon). Finally, "to domesticate" can mean "to tame or bring under control." To tame: "to reclaim from the wild state," "to overcome the wildness or fierceness of (a man or an animal)," "to subdue, subjugate," and "to bring under the service of man." Added to the notion of control, we see here the common superstition of Euro-American thought, that the human realm is split from the so-called "wilderness." Anything human is not wild, and the places that humans inhabit are no longer wild.

If our goal is to find a definition of *domestication* that works in the context of salmon, we must choose our steps carefully. Our definition must be clear enough to distinguish domesticated salmon from undomesticated salmon while not burdening itself with too much proximity to ambiguous concepts such as wilderness or nature. And although there does seem to be a widespread connotation of the verb with control and with power, I agree with Lien and Law that the definition ought to avoid deciding that there is an imbalance of power in advance. It will be more useful to pose the issue of power as a problem: Rather than settling the issue beforehand, our definition should invite us to question power relations as they are enacted through concrete practices.

My working definition of *domestication* would be as follows: "To domesticate is to gather in or around the nearness of people, and to make the close relationship more or less permanent." The definition allows us to ask: Is domestication a relationship that is characterized by uneven power relations? Is it a mutually beneficial relationship? It depends. It depends on circumstance, and it depends on those who are doing it.

The writer Lierre Keith recalls the precise moment she realized that the relationship she has with her chickens was not about control or domination. It was "six o'clock on a January morning, and well below zero," she recalls.[13] She was carrying a gallon of hot water through ice-slick snow. The day before some of the water had dripped into the doorjamb and it had frozen rock solid. She brought out screwdrivers, butter knives, and matches, and while she was sensing an increasing pain in her palm and was showered with snow that fell from a roof, it suddenly hit her: The chickens were warm and fed and safe, and she was miserable on their behalf. "That wee drip sliding down my spine was like a cold jab of reality," she recalls. "Chickens have gotten humans to work for them. In exchange, they take care of us, but not by bringing us water. By providing food—meat and eggs—and a whole constellation of other activities useful for farms. It's a partnership, and one that worked out well for both parties until factory-farming."[14]

Michael Pollan also suggests that we need not necessarily think of domestication as something humans do to other creatures. We can think of it in reverse: something other animals or plants do to humans. Pollan writes: "The species that have spent the last ten thousand or so years fig-uring out how best to feed, heal, clothe, intoxicate, and otherwise delight us have made themselves some of nature's greatest success stories."[15] Compare the fifty million dogs in present-day America with the roughly ten thousands wolves, and you get the picture. "So what does the dog know about getting along in this world that its wild ancestor doesn't? The big thing the dog knows about . . . is us: our needs and desires, our emotions and values, all of it has folded into its genes as part of a sophis-ticated strategy for survival. If you could read the genome of the dog like a book, you would learn a great deal about who we are and what makes us tick."[16] And what is true of dogs is true of certain plants as well, Pollan suggests. When it comes to the apple, the tulip, cannabis, or the tomato, "We could read volumes about ourselves in their pages, in the ingenious sets of instructions they've developed for turning people into bees."[17]

Domestication might become a relationship of domination, but it need not necessarily be so. What is commonly seen as one-way exploitation might just as well be a relationship of mutual benefit and coinhabitation, a relationship closer to symbiosis than to parasitism. Domestication might describe a relationship that enacts reciprocity as its fundamental ethical orientation. The question is, is it reciprocal for feedlot salmon?

To Eat or Not to Eat

Lien and Law insist that feedlot salmon have some degree of agency because they might choose either to eat or not to eat. But the salmon's choice to eat or not to eat exists entirely within the constraints set up for them by their keepers. Humans decide the circumstances under which the fish will be fertilized and grow, what they will be fed, where and how they will die, and who receives their flesh afterward. It is strictly within these non-negotiable boundaries that salmon can "choose" to either eat or not. To call this "agency" would be the same as saying that factory-farmed chickens possess agency, for they, too, are presented with the same Shakespearian question "to eat or not to eat" (and, like farmed salmon, they are given practically to nothing else).

If this is agency, then it is a limited one. It is not the agency to assume responsibility for their own lives. It is not the agency to eat when, where, and whom or what they choose. It is not the agency to follow the migration patterns of their ancestors, nor to renew their intimacy with the watershed community after every oceangoing journey. It is not the agency to do their dance when they know the time is right, nor to die when they have spawned, nor to return their bodies to the land. None of these more relational and complex aspects of salmon agency are available to salmon in feedlots. Their lives and deaths are confined strictly by the infrastructure set up around them by human decision making. Pens clearly define an inside away from an outside. Pellets leave the salmon with no alternative but to eat a semivegetarian diet.[18] Assembly-line slaughter, filleting, and packaging take the fish further away from the land and deprive the salmon of the possibility to gift themselves back to it.[19] But on top of all that: Atlantic salmon are bred to be constantly hungry! Their sole choice, "to eat or not to eat," is being gradually bred out of the population.

The anthropologist Anne Magnussøn, a student of Lien's, spent her PhD research studying every part of the Norwegian salmon industry's consumption chain, from fertilization to growth to slaughter and processing. To do so, she visited several inland smolt-rearing sites, the places where juvenile fish are hatched and raised until they are ready to undergo the transition from fresh water to salt water. Typically, these sites keep the juvenile fish separated by developmental phases. Alevin—just-hatched salmon—are kept together in small tanks, fry are kept together in larger tanks, and so on. During a tour of the premises, Magnussøn spoke with Bjørn Erling, the manager of the site, and learned that all smolt-rearing sites face a common problem: Salmon are inherently social creatures.

Salmon strive to establish a social structure even in a confined and vacant place like a featureless plastic tank, and these social structures will affect their growth rates. If left undisturbed, they will not all develop in sync with one another. Some will grow larger and some will remain smaller. This is a problem for those who have an economic stake in the young fishes' growth. Since biomass is the imperative around which the salmon industry's infrastructure and rearing processes are designed, smolt producers have thought of an ingenious solution to the problem of salmon sociality—they simply have to be sorted and moved to another tank: "When the fish were moved and graded by a machine according to size, new hierarchical orderings had to be established. *Moving fish and upsetting their social order was necessary according to Bjørn Erling. . . . This situation is rectified in the rearing house by regular grading through the use of a machine.*"[20]

In other words, juvenile fish—sentient beings with an inert striving to relate socially to other sentient beings—are being explicitly, systematically, and forcibly shaped into conformity. The imperative of uniform growth requires their striving to establish a social order to be constantly interrupted. They must be treated as manageable objects because to treat them in any other way imperils a reliable, steady, and even production of similar and calculable units. The assumption that salmon are biomass has direct impacts on their lives as individuals and as a collective. Agency is irrelevant, assuming we understand agency as the ability to engage creatively in the unfolding of every aspect of their lives both as individuals and as a collective. In industrial salmon production, there is no place for such agency. Salmon, as seen by the industry, constitute a vast and steady river that flows from feed to flesh to cash, and this river must not, under any circumstances, be obstructed.

Elusive Salmon

Lien and Law further point out that the salmon's "slippery nature" in the holding pens helps to ensure the reciprocal relationship between humans and salmon. Submerged in water, the fish are not permanently subject to the human gaze. They remain elusive.

The individual fish might remain elusive, and their minute-by-minute whereabouts might be hidden from view. But these are negligible details to a production process whose focus is elsewhere. Lien and Law themselves write that the industry is explicitly *not* interested in knowing individual salmon, and Magnusson's work shows that salmon are actively intercepted in their attempts to form social bonds. Biomass remains the central ontological category. As biomass, salmon are in no way slippery; they are subject to near-total control.

Lien and Law also ignore the fact that salmon production involves both an inland freshwater phase and an ocean-pen saltwater phase. Their argument about the elusiveness of salmon focuses only on the saltwater phase, but feedlot salmon spend as much as half their lives in inland containers. As Anne Magnussøn has observed, the freshwater phase is characterized by a "wish for absolute control with the fry's built environment."[21] No elusiveness there.

For the salmon, this wish for absolute control also means that they are becoming increasingly dependent on the infrastructure. If that infrastructure suffers a blow, they will face the consequences. Lien and Law claim that "[f]armed Atlantic salmon do not necessarily depend on humans for survival,"[22] but if we examine the entire rearing process, this is plainly incorrect. As Magnussøn reflects: "After having walked through several rooms with fish tanks of different sizes with Bjørn Erling, we reached the room with the aggregate. It was checked several times a month. . . . I had not reflected on the electricity's vital role in the rearing unit before I was shown the huge aggregate. . . . A power failure, where the aggregate failed to kick in, would kill all the fish."[23]

Wild salmon are also elusive in ways that farmed salmon are not, in that the salmon life cycle is subject to seasonal fluctuations. Seasonal fluctuations in the animals' growth patterns would ultimately be reflected in fluctuations in productivity—there are certain times of year when you can't buy fresh wild salmon. This is a problem if the animal is regarded only as biomass. Farm-raised salmon, as an industry driven by market capitalism, must try to conform to the demand of "continuous production," as Magnussøn writes.[24]

How do you accommodate an animal to market capitalism when that animal is adapted to seasonal fluctuations of light and temperature? Manipulate the seasons!

At the end of their freshwater phase, salmon undergo a smolting process in which they prepare to live in salt water, a drastically different medium. Some changes, such as the loss of parr marks or the change in camouflage, are visible to an outsider; others, such as a changed metabolism that will enable them to survive the more saline conditions of ocean water, are not. Smolting is triggered by a combination of light and temperature. If you manipulate both, you can get the young fish to smolt whenever you require. As Magnussøn writes: "If one wants the fry to smolt in autumn, light is applied all day and night until summer. After this an artificial winter in terms of lighting is devised. This will trigger smolting in the autumn. When one wants the fish to smolt in the spring, one uses 24 hour lighting until September. After this phase, the fry is given light adjusted to the hours of light outside. This fish will then smolt in the spring."[25] Artificial seasons are one of the reasons consumers can now purchase salmon throughout the year.

Who are salmon here? They are supply to retailers and customers. They are coherent units, in tune not with the flux of seasons but with market forces. They are a commodity, perfectly accommodated to a notion of reality that is centered on the human. This is a power relation barring any significant degree of reciprocity.

Escapees

The first recorded attempts by humans to reproduce salmon occurred in France around the year 1400.[26] The modern salmon industry is the most recent expression of a long and complex historical involvement between humans and fish, and surely, there does seem to exist a gradation of "more-or-less domesticated" fish. Lien and Law accurately observe that the boundaries between domesticated and undomesticated salmon can be fluid.

On one end of the spectrum are the free-ranging and clearly undomesticated salmon, self-reliant populations strong enough to survive without human interference. Such populations are rare today, restricted to their remotest and northernmost ranges of North America, Scandinavia, and Russia. Then there are free-ranging and partially domesticated salmon, those who complete their entire life cycle at sea, but whose numbers have fallen so low that they would be unable to sustain their runs without the contribution of hatcheries. There are also salmon runs that have collapsed entirely, but that are being kept artificially alive in tanks for future re-colonization of their former home. The Vossolaks are an example of these. The only "real" Vossolaks left by the year 2012 were offspring of salmon taken from the Vossoelva in the wake of the notorious collapse in 1987. Between World War Two and 1987, the Vosso watershed had seen annual commercial catches of nearly ten tons, but a combination of human-made calamities caused the stock to collapse completely. Hydropower development, road development, acid rain, and the building of the Nordhordaland Bridge, as well as farmed escapees and sea lice, all contributed to diminishing what might once have been, according to Bjørn Barlaup of the University of Bergen, the most majestic salmon population in the North Atlantic.[27] Some Vossolaks have lived on in their aquariums, safeguarding the genetic memory of their local lineage. They represent the germ cell for an ambitious repopulation program that has been trying to give back to Vosso its own, native fish.

Between these salmon, the boundaries are porous. The distinction between domesticated and undomesticated salmon is not always clear, nor is it always meaningful. Vossolaks in aquariums fit my definition of domestication, and yet they are kept as domesticates precisely so that their lineage will one day become once again self-reliant, self-sustaining—undomesticated. This logical paradox shows that the dynamic of domestication is nonlinear and reversible, and that domestication can be an attempt to repair losses suffered earlier.[28]

On the far end of the spectrum are the fully domesticated salmon, the feedlot salmon. They eat what they are given. They die when it is decided upon. Pens clearly define an inside away from an outside. But Lien and Law remind us of an important consideration: Feedlot salmon can escape from their pens! Escapees seem to swim boldly across theoretical margins, and their numbers are large. In 2010, the salmon who escaped from Norwegian farms outnumbered the free-ranging salmon who returned to Norway's coast that year to spawn.

So how should we think of escapees? The problem of escapees, as described by the researchers Herve Migaud, John Taylor, and Tom Hansen, is as follows: No measure of precaution can guarantee that "reproductively competent" salmon in open-sea containers will not find a way to escape their confinement, and when they do, nothing can prevent them from living out their "competence" beyond the reach of human control.[29]

If fertile escapees threaten undomesticated fish, the solution is to produce infertile salmon. Several methods are being tried out to sterilize salmon, among them a method called triploid induction. Triploid induction is the artificial creation of fish with not two, but three sets of chromosomes, making them sterile. If triploid fish escape, they live and die without ever producing offspring. This protects the wild salmon's genes from being contaminated by feedlot escapees. But triploid salmon also have benefits for investors: Triploid salmon grow faster, have greater harvest windows, and require fewer phases of the rearing process, cutting down on running costs. Because the triploids are genetically modified, investors can even protect their intellectual property rights by patenting them.[30] It's a win-win situation for both wild salmon and investors.

There is a third winner as well: the researchers. At the time Migaud and his colleagues published their paper, their knowledge base for triploid salmon was still insubstantial. Their technology was all conceptual. They had the ideas but lacked funding. It was against this background that they wrote their list of benefits to triploid induction. And it was against this background that they mentioned, in passing, that their project was still "under negociation [sic] to finalise funding."[31]

They did finalize funding.[32] The result was a win-win-*win* situation, with free-ranging salmon, the industry, and the researchers equally profiting from the arrangement.

But there is a fourth party to consider, though it receives no mention in Migaud and his colleagues' discussion. It is not entirely clear what benefits, if any, the contained salmon get from being modified into triploids. The question of whether they benefit is not asked, suggesting that it is not worth considering. Anthropocentrism's narrative lens makes certain actors and problems visible and conceals others. Contained salmon are objects, *res extensa*, there to be plasticized into whichever form or content will benefit those whose considerations must come first. Concerns for their welfare are subordinate, functions of the overarching imperative of containing salmon in feedlots.

Sterilizing salmon is an efficient and reliable measure to contain your product, assuming you can automate the procedure and your narrative blinder is strong enough to let you avoid the question of whether sterilizing benefits those who are sterilized. But moving salmon feedlots into closed inland facilities is even more effective at containing those who must not get away.

The industry had long been reluctant to pursue this thought. Inland facilities were too difficult to build, some argued. They were not cost-effective and would leave Norwegian producers unable to offer a competitive product in an expanding world-market, others said. This made inland facilities difficult to champion in an industry with such strong national ties as the Norwegian salmon industry: If you can build closed-containment feedlots, you can build them virtually anywhere, and that means Norwegian manufacturers would lose the competitive edge that the country's long and rugged coastline had previously given them.

It might not come as a surprise, therefore, that the first successes came from elsewhere. In February 2012, the Conservation Fund Freshwater Institute and the Atlantic Salmon Federation issued a press release stating that they were pleased in every way with the salmon they were farming in Shepherdstown, West Virginia, some 100 kilometers from Chesapeake Bay. These fully contained salmon were healthy, grew well, and had received encouraging reviews at taste tests in Manhattan and St. Andrews.[33] Bill Taylor, president of the Atlantic Salmon Federation, said that "the trial is proving that we don't need the ocean to produce farmed Atlantic salmon for market."[34] Inland pioneers slowly scaled up their efforts: The Conservation Fund Freshwater Institute has already produced more than 200 tons of trout, char, and salmon, with numbers on the rise. Bill Taylor expressed his conviction that their trials would constitute a precedent.

It seems he was right. The company Billund Aquaculture is now working on a joint Danish-Chinese endeavor to build salmon tanks in the Mongolian desert, one of the remotest places anywhere, far from any ocean. Bjarne Hald Olsen, a leading engineer of the project, puts it as follows: "As long as you have access to water and electricity, you can produce salmon anywhere you please in the world."[35] In Norway, the debate around moving salmon inland continues to draw heated responses both from fervent proponents and from impassioned opponents. The former minister of fisheries, Lisbeth Berg-Hansen, called the trend "unrealistic" and "problematic"[36]—statements that must be considered in light of Berg-Hansen's own co-ownership of feedlots. Her predecessor as minister of fisheries, Svein Munkejord, has claimed to see an imminent sea change in the salmon industry. He has invested several hundred thousand kroner of private funds in a project seeking to move salmon inland—in Ireland. He justified his decision with a remarkable nod to ecological concern: "By investing in land-based farming, *environmental arguments* [such as the industry's notorious issues with escapees, lice, or eutrophication] will fall away. Such structures will *give full control*."[37] Creditable though this "environmental" concern might be, Munkejord's ideas also speak of the difficulty to accurately articulate such concern in ways that fully resonate with ecological thought, free from the constraints imposed by the story of human exceptionalism.

AquaBounty, too, has some solutions to offer for the problem of escapees. It has adopted a zero-tolerance policy for escapees, envisioning a production cycle so hermetically sealed off that no genetically modified salmon will ever be given the chance to interbreed with free-ranging salmon. Knowing full well that there are no individual measures that can provide a 100 percent escape guarantee, the company has concocted a safety net of overlapping and partially redundant measures to contain their AquAdvantage salmon.

There are, first, biological measures. All the fish produced by the company are female, made possible through a breeding strategy that is considered 100 percent effective. All of these females will also be treated to possess not two but three sets of chromosomes in their cells. Like other triploid creatures—creatures with three sets of chromosomes—the resulting fish will be sterile.[38] These biological measures are meant to prevent cross-breeding between GM salmon, as well as make breeding of female GM salmon with non-GM male salmon nearly impossible.

Then there are physical measures to contain the fish. All AquAdvantage salmon will be bred exclusively in inland sites. They will spend their juvenile phase in tanks on Prince Edward Island, Canada. Smolts will then be shipped to another inland facility in Panama, where they will grow until they are ready for slaughter. Along every step of the process, tanks, screens, filters, covers, and nets will be installed to prevent their escape. The company mentions also that in the unlikely case of uncontrolled escapes, chemicals such as chlorine are available to kill the fish. It also plans to implement strict security routines. These will include preventing unauthorized personnel from accessing the production site, surveying the movements of those who are authorized to be there, and preventing predators from entering. To make sure that neither unwanted humans nor any other animals will be able to come anywhere near the tanks, the company plans to install fencing topped with barbed wire around the entire facility. All in all, the company will erect eleven sequential barriers between the fish in the tanks and the nearest river.

Finally, biological and physical containment measures are complemented by measures of geographical containment. The grow-out site in Panama is situated in a tropical environment, and the nearby river drains into the Pacific at a latitude thought to be inhospitable to any salmon. Water temperatures in the lower reaches of the watershed are so high that in the unlikely event that any GM salmon should ever make it that far alive, the heat stress would kill her before she could reach the ocean.[39]

The containment does not end with the living fish. Salmon that die before they are ready to be slaughtered will be disposed of in a controlled manner. Those that die on Prince Edward Island will be frozen, collected, and eventually incinerated together. Those that die in Panama will not be burned but will be buried on-site. The company has laid out precise routines for how these burials are to be organized.[40]

These trials attempt to escalate control and to close any still-permeable pores in the boundary between wild and domesticated salmon. The demarcation line might never have been absolute, but the direction is evident: to solidify the domestication process and exercise full control over the feedlot salmon.

Despite all these measures, there remains at least the theoretical possibility that some fertile salmon will get away. And when they do, they continue to disregard our attempts to conceptualize them. Escapees have a chance to survive without humans. They might pass on their own genetic memory. They might become free-rangers.

Escapees impact other fish in a number of ways, none of which are beneficial to the wild salmon. In a special report to the Norwegian Institute of Nature Research, Eva Thorstad and her colleagues write that "inter-breeding of farmed with wild salmon can result in reduced lifetime success, lowered individual fitness, and decreases in production over at least two generations."[41] Hybrids compete with wild salmon for food, territory, and spawning grounds. They might transmit diseases and sea lice. They often lay their eggs later than the wild fish, displacing eggs buried in suitable locations before they arrived. Thorstad and her colleagues also describe the competitive advantage that domesticated juveniles have over wild salmon: "Farmed juveniles and hybrids are generally more aggressive and consume similar resources as their wild counterparts. In addition, they grow faster than wild fish."[42]

Genetic differences also impact the wild salmon population in far-reaching ways. Free-range fish have always been buffered against periodic calamities by their broad genetic repertoire. With their large and diverse genetic base, they have survived every upheaval of the two northern oceans for millions of years. Compared to their wild cousins, domesticated salmon are genetically impoverished. Nearly all hybrids in use today were originally crossed from fish taken from only forty Norwegian rivers.[43] Breeding has concentrated predominantly on producing hybrids able to satisfy mounting economic interests. Salmon have been bred for faster growth, increased body size, greater stress tolerance, delayed maturity, greater temperature tolerance, flesh quality, disease resistance, and more efficient egg production.[44] They have adapted to thrive in the habitat assigned to them: feedlots.

Biomass remains the index of their performance, the central node where the various activities, intentions, and actors run together. Measured by that index, hybrid salmon are one continuous success story. In 1980, the industry in Norway produced 4,312 tons of salmon. In 2010, annual production exceeded 1 million tons for the first time ever. But when feedlot salmon break free, they carry their indistinct genetic stew into populations highly specialized, highly adapted, and highly localized. When they find free-ranging mates, it is not surprising who suffers: "Hybridization of farmed with wild salmon . . . has the potential to genetically alter populations, reduce local adaptation and negatively affect population viability and character." Large-scale, whole-river experiments in both Ireland and Norway "found highly reduced survival and lifetime success of farmed and hybrid salmon compared to wild salmon."[45]

This implies that the industry objectifies not only the domesticated salmon but also the free-ranging fish. Although large-scale impacts on wild populations are well documented, these losses do not instigate any sustained, fundamental debate on the industry. For the time being, the losses are accepted as inconvenient externalities to the manufacturing process. The process itself must not be questioned. Wild salmon, like farmed salmon, become functions in the industrial framework. Both are being defined into *res extensa*. They do not have intrinsic value. Concern for them has no place in the framework.

Exploited Captives

In her book *Fiskenes Ukjente Liv* (*The Unknown Lives of Fish*), veterinarian and author Bergljot Børresen describes the current wave of industrial animal husbandry, beginning in the 1950s, as the second major revolution in the relationship between human and other-than-human animals, comparable in its magnitude with the first, which began in the wake of the last ice age. A substantial difference between this current industry-driven revolution and earlier forms of animal husbandry, Børresen suggests, is the degree of reciprocity between humans and their domesticated companions. The first domestication wave some ten thousand years ago was marked by high degrees of reciprocity. Humans offered themselves as extended "emotional family" to the boars, goats, wolves, or chickens alongside whom they gradually came to live.[46] Their "predatory nonemotionality" vis-à-vis these former animals of prey was largely suspended, as humans and animals "greeted" or "acknowledged" one another as peers in newly emerged, cross-species social groups.[47]

Børresen also shows that direct, personal encounters were a critical factor for the emergence of such cross-species social groups. Both human and other-than-human animals greeted one another, each in their distinct manner. Humans spontaneously responded to this cross-species communication with empathy, and they could see their animal companions in their individuality. With their predatory nonemotionality on hold, humans assumed the roles of caretakers, guardians.[48] Strong social bonds were formed that endured across several generations. Domestication would have been a *symbiotic* process, much like Sandor Bökönyi has described it: a process of gradually learning to live together for the mutual benefit of all.[49]

The salmon industry's ideal, by contrast, is to encounter the fish "objectively" through the "disinterested" lens of science. This ideal of analytical distance is coupled with a strong imperative for economic gain. In combination, the objectifying stance and economic reasoning lead to what Børresen calls a state of "chronic nonemotionality":[50] the permanent shutting off of any experiences of empathy with the living creature. Modern industrial practices obstruct possibilities to encounter other animals in their individuality, to form robust social bonds with them, or to develop social practices that would strengthen such bonds further. They create a perceptual lifeworld in which feelings of empathy, compassion, even love for another being, are rendered redundant.

Of course, there are still some personal encounters, some moments of emotional bonding between humans and feedlot salmon. Think of the feedlot manager who does daily hand-feeding rounds amongst "his" fish. But does this lone feedlot manager change the fact that feedlot salmon are categorically viewed as biomass?

No, but it does show us that even within the context of the grossly imbalanced, nonreciprocal salmon feedlot industry, some people still strive for contact! Making contact seems to be an inherent need for us as highly social animals, and some of us will continue to form emotional bonds with other animals even under the least favorable conditions. Even when the industry's categorical imperative—to redefine the living creature as biomass—makes itself felt in concrete practices, in infrastructure, and in speech, some practitioners will still create small pockets within this larger feedlot cosmology where they can nurture that resilient impulse for contact.

At least part of the success of modern feedlot practices might be their ability to absorb such countertendencies within their larger cosmology. Feedlots do provide very limited space to forge emotional bonds with animals. Those bonds might not be reciprocal, and they might be only a faint echo of what could be experienced in less quantity-driven, less intensive, and more personal farming practices. But they might be just strong enough for some practitioners to satisfy the most basic need for contact. This makes industrial feedlot farming a near-perfect, near-invulnerable cosmology. The innate human propensity for personal and emotionally charged bonds with other animals has the potential to subvert and destabilize the industry. If the dissatisfaction were to reach a certain critical threshold, it could trigger an outpouring of resentment. But because that subversive potential is being absorbed into the industry's larger cosmology, all wind is taken out of its sails. Even as the industry stymies deeper and more nourishing encounters between humans and salmon, it remains impermeable against attempts to pierce through its perceptual stronghold.

The salmon industry has succeeded in nearly perfecting a principle inherent to the story of human exceptionalism since the days of Descartes: To exploit another creature, one must actively and artificially create a sharp rift between "us" and "them." One must create an Other which is no longer subject to "our" normal ethical concerns. Descartes was first to recognize a crucial aspect of this Otherizing: To create an efficient separation between "us" and "them," one must actively suppress one's spontaneous empathy with the suffering of these other animals! This was the lesson Descartes taught his students when he urged them to ignore the screams of the vivisected animals. He must have understood that the unlearning involves action. The impulse to empathize with the suffering of other sentient beings is strong in humans.

The poet and salmon restoration activist Freeman House writes that empathy "comes from a Latin word meaning to suffer with gathered senses. . . . Empathy with lives that are alien to our own is the human impulse that gives rise to vernacular practices that celebrate and regulate our links to other species."[51] Empathy is an impulse, a spontaneous experience. It is an impulse that helps us celebrate and regulate our encounters with other beings. It is also a suffering *with gathered senses*. Empathy, in other words, requires an overcoming of the Cartesian split: We must reflect on our encounters with other beings not only with reason, but with all faculties of knowing, including feelings, intuition, and the senses. "Gathering" here is not so much "collecting," but rather reintegrating, un-scattering, or, in a rather direct sense, incorporating. Empathy presupposes embodiment.

The salmon industry recruits a richly layered semantics of separation, scripting the encounter between humans and salmon through the disinterested, alienating language of the rational intellect. Salmon are "resource," "product," "brand," "biomass." Such language appeals to the Cartesian ideal of a disembodied *res cogitans*, a thinking mind that is wholly separate not only from the world, but from its own body. Such language consistently and systematically excludes other ways of knowing—intuition, feeling, and sensing. The thinking mind alone is brought to the encounter. It alone defines what ways of speech are appropriate; it alone defines the truth-criteria. The spontaneous impulse that is empathy becomes scattered in the process.

The scattering of this spontaneous impulse holds salmon apart as an Other with a cold, objectivity rationality, but it also un-gathers the human senses. The Cartesian split is not, first and foremost, a split between humans and other-than-human beings. It is a split right through the human. Descartes applied his scalpel not only to the soft, warm bellies of the vivisected animals. He also applied another sort of scalpel to his students: They, too, left the vivisection lessons sliced open, partitioned, wounded. Their psyche had been cleanly amputated from their sensing, feeling, intuiting bodies.

These complementary assaults are mobilized by the salmon feedlot industry. It objectifies the salmon while idealizing one human way of knowing at the expense of all others. Humans are not allowed to participate in the encounter with their full being. They are prevented from entering the relationship as embodied animals with intuitions, feelings, and a richly textured sensual apparatus. Humans, like the salmon, are being demeaned, scattered, reduced. The narrative of anthropocentrism is also the narrative that suppresses those aspects of the human that are perceived as "animal-like": feelings, intuitions, and the body's exquisite, marvelous senses.

The salmon industry is not unique among feedlot industries, which all practice such a reduction of other animals and of humans. The same reductionist tendencies script the practices and discourses of cattle, hog, and poultry production. But there are two unique characteristics that add to the perceived invulnerability of the salmon industry.

If you do a quick internet search of pictures of salmon feedlot sites, you will find that they are uncannily beautiful. Round pens are set against stunning backgrounds, perfect geometrical shapes that float on a calm ocean surface. They might be bathed in radiant Arctic light, or there might be snow-covered peaks in the far distance or lush green cliffs that drop precipitously into the ocean. I visited a farm outside the Lofoten archipelago in northern Norway, and in my body, I still hold a vivid recollection of the place: the soothing quiet of the ocean, the enormous vault of the Arctic sky overhead, the unspeakable beauty of distant mountains, the killer whales who came close to the pens, the sight of a golden eagle who soared directly above our boat. The setting and its infrastructure render the suffering of the sentient animals who are being held in captivity all but invisible. Even when we look directly at the feedlot, we see beauty! Being embodied minds, we do not respond to what we see as rational, purely objective thinkers. The tranquil view of perfectly round rings floating in Norway's magnificent landscapes appeals evocatively to our sense of beauty, to feelings. Even when we know about the systematic abuses and the suffering, the scene still doesn't quite feel like a site of concentrated cruelty. This disconnect between knowing and seeing works in the industry's favor.[52]

Unlike other domesticated mammals or birds, salmon are biologically that much more remote from us. They might be social creatures, but even if we did encounter them as embodied beings, face-to-face, we might still find it hard to empathize with them. As Børresen writes: "Few people experience spontaneous, embodied empathy with fish. Nearly nothing in the fishes' looks and behavior acts as a social trigger for humans. Fish have an extremely low 'Bambi factor'. . . . Individual encounters in which fish and human acknowledge one another are practically nonexistent." Børresen does qualify this, pointing to countless incidences in which divers have encountered curious fish who have sought their nearness; some fish even appear to recognize individual divers from earlier encounters. But, she says, "The large production units offer no room for such phenomena."[53]

The perceptual gap between our species is that much wider; a spontaneous sense of connection, or even cross-species communication, is that much harder. Salmon live inside a fluid, three-dimensional medium, so unlike the soil beneath our own animal feet. The salmon's lifeworld might be a region of the larger biosphere that we inhabit, but theirs is a remote region, one to which we do not easily travel. This makes it that much harder to assume what Børresen calls an *ichthyocentric* viewpoint, to experience the world from a fish point of view.[54] It makes it that much harder to think, feel, intuit, and sense our way into a deeper, empathetic understanding of who salmon are, and what it is like to be them.

Those last two questions—who salmon are and what it's like to be them—will continue to occupy us. In this chapter, we have simply sought to understand who salmon are within the cosmology of the feedlot industry. Børresen suggests thinking of animals bred, raised, and killed in industrial feedlots as "exploited captives."[55] She borrows the phrase from Juliet Clutton-Brock, and it is a powerful phrase. *Exploit*, according to the *Oxford English Dictionary*, means: "To 'work' (a mine, etc.); to turn to industrial account (natural resources) b. *transf.* To utilize for one's own ends, treat selfishly as mere workable material (persons, etc.); to 'make capital out of'."[56] Intuitively and by definition, *exploitation* is a strong word, but it does precisely characterize the current industrial practices around these living, sentient beings. Feedlot salmon are not "tamed pets,"[57] nor are they "farmed animals."[58] Their agency within the pens is so restricted that invoking the concept of agency grossly misrepresents the power relations between the fish and those who build, use, and conceptualize the infrastructure around the fish. Børresen's suggestion is justified: Feedlot salmon are exploited captives.

As this observation sinks in, it will begin to fester like a sore, making the story of anthropocentrism that much harder to maintain.[59]

Gathering Our Senses

A more fundamental critique of the cosmology enacted by the feedlot industry is required. We must look more deeply at the geographically concrete, historically contingent, and nonlinear human-salmon relationship, and seek possibilities for mutually enriching encounters across the species divide. How can we speak a language that is receptive to the salmon's other-than-human eloquence? How can we pierce through the perceptual wall of feedlot cosmology?

That perceptual wall appears so unassailable and solid because it is reinforced several times over: through practice, through infrastructure, through speech. All of these aspects feed the narrative of separation, supercharging it with positive mythical force. The industry appears desirable, successful, even worth identifying with. But the observation remains that it systematically exploits hundreds of millions of sentient beings in Norway, a country frequently cited as one of the world's happiest.

The hope for change might lie with something the philosopher Charles Eisenstein has written: "Any action that is open to symbolic interpretation can be part of the telling of a story. And that is every action."[60] Any action, rightly interpreted, can open a rupture in a story that seems impregnable.[61] Once that first rupture opens, there is a pressure point where one can articulate a sustained critique.

The narrative of separation as enacted by the salmon feedlot industry needs to actively counteract empathy with other-than-human living beings. Empathy is a spontaneous experience that forcefully shatters any notions of being fully separate. This is our pressure point. If the artificial separation between "us" and "them" requires us to actively unlearn empathy with other animals, then to mend the rift, we must practice the reverse strategy: to actively create the conditions for empathy to make itself felt again.

The narrative of separation, of anthropocentrism, is a narrative that suppresses those aspects of the human that are perceived as animal-like. Having empathy with other-than-human animals means allowing ourselves to reconnect with those animal-like parts of ourselves. Salmon make themselves more fully known not only to the degree that we learn more about these enigmatic creatures out there in the oceans, but also to the degree that we gather our own senses. In this way, we can begin to move beyond the Cartesian split.

CHAPTER FOUR

Keystone

Enriching one another through the sharing of our
existences becomes a communal discipline.

CATHERINE KELLER

S
ome of the highest tides in the world sweep these shores. In Kvichak Bay or Nushagak Bay, the tidal extremes can exceed ten meters. Winds are strong, and shoals, sandbars, and shallows make seafaring a challenge in a place that, not long ago, was a dry and lush part of the Bering Sea Land Bridge. That bridge stretched from Eastern Siberia to Northwest Alaska, connecting the Eurasian and American continents and enabling bands of wise apes, among others, to spread from west and to east, moving into a double-continent that until about twelve thousand years ago had never known anyone like them. Twelve thousand years—that is between four and five hundred generations. This is how many human ancestors have been born from and laid back into the land that is the Americas.

The other participants in this story have a lineage that extends far deeper into the continent's memory. The oldest known ancestor of all modern salmon, *Eosalmo driftwoodensis*, lived in the lakes of Western Canada between forty and fifty million years ago. Between ten and twenty million years ago, during the Miocene epoch, global cooling formed an ice sheet in the Atlantic polar regions that prevented the salmon populations on either side from interbreeding. The salmon speciated into two different genera, the Atlantic *Salmo* and the Pacific *Oncorhynchus*. Ten million years ago, the Pacific salmon began branching out into clans, and by around six million years ago, during the Pliocene epoch, these clans had speciated into eleven distinct Pacific species—on the Asian side, the Northwest Pacific masu salmon (*Oncorhynchus masou*) and the red-spotted masu, or amago (*O. masou macrostomus*), and on the North American side, the Chinook salmon (*O. tschawytscha*), the pink salmon (*O. gorbuscha*), the coho salmon (*O. kisutsh*), the chum salmon (*O. keta*), the sockeye salmon (*O. nerka*), the kokanee (a subspecies of the sockeye who lives only in landlocked lakes), as well as cutthroat, golden, and Apache trout (*O. clarkii, O. mykiss aguabonita, O. apache*). There was once a twelfth as well, sabretooth salmon (*O. rastrosus*), a fish who could weigh up to 350 pounds, roughly the weight of an adult panda, but this giant slipped into extinction long ago. Because salmon have very different life spans, six million years will be anywhere between seven hundred fifty thousand and three million generations. And that is just the species we know today. Include the distant ancestors who lived in the waters of Western Canada, and you can figure that anywhere between six and twenty-five million generations were born from these rivers and returned to them at the close of their lives.

These numbers are so large they become easily intangible. But they are a shorthand for a reciprocal metamorphosis between bodies and land. They point to a participatory shapeshifting that takes form in the miniscule, translucent red globes of salmon roe, or that stalks the forest ground as grizzly, or that grows down into that same ground as the limbed, wooden body of old-growth spruce. Before and beyond each of these overlapping, individual life cycles, there is the annual pulse of the great salmon run, which enriches watershed communities through an ever-renewing infusion of vital nutrients.

The philosopher J. Baird Callicott has written that both the new physics and ecology are now emerging into a "consolidated metaphysical consensus . . . which may at last supplant the metaphysical consensus distilled from the scientific paradigm of the seventeenth century."[1] The core idea of that consensus is that "we are enfolded, involved, and engaged within the living, terrestrial environment—that is, implicated in and implied by it."[2] Like the new physics, the science of ecology has begun to usher in a turnaround in our understanding of Being, including a re-evaluation of our place as two-legged, inquisitive mammals in relation to the larger-than-human world.

This change is a wholehearted departure from dominant Western metaphysics, a departure that suggests that the monocentric narrative of human exceptionalism is dangerously anachronistic. Narratives have a certain inherent inertia. They need not necessarily be in sync with philosophical or scientific breakthroughs. It is possible to think of the salmon industry as one such anachronism: It enacts a cosmology that is quickly falling out of touch with our maturing ecological understanding, while also understanding itself as the future of our human-salmon relationship. One reason for this discord might be, as Callicott observes, that the ecological turn brought about by contemporary holistic science implies a way of thinking that is "virtually unprecedented in the Western intellectual canon."[3] The human-centered narrative tradition has had very little training in holistic thinking. Instead, it continues a two-and-a-half-thousand-year-old track record of cultivating an atomistic and materialistic thinking that still subtly influences the stories we tell.

Salmon are a telling creature to turn toward as we discuss the ontological revision that the science of ecology is helping to bring about. Accumulating evidence has recently drawn an increasingly detailed and sophisticated tableau of salmon as "keystone species." The term is borrowed from the craft of masonry. A keystone is that uppermost stone in an arch of bricks, the central one that holds the entire construction in place. In its absence, gravity will bring the arch to a rapid collapse. Keystone species hold a place of similar significance for the ecological communities in which they live. When they are present, the community becomes richer, more diversified, and more resilient. When they are absent, the opposite is true: Their disappearance sends ripples through the ecological web, resulting in less diversity, less resilience, and greater impoverishment.

Picture thirty million sentient beings migrating from the deep sea to the shallower waters of the bay, all moving in the same direction, all sensing, in their conscious bodies, the magnetic pull of a place they once knew. They have spent their adult lives at sea, feeding on shrimp and other smaller oceanic creatures, as well as zooplankton, which they strain from the water with their gill rakers. Salmon gain over 90 percent of their weight at sea. They carry vast quantities of marine-origin nutrients in their bodies when they return to their natal streams to spawn and to die. Their bodies are living reservoirs of protein, fats, and nutrients like nitrogen and phosphorus. When these bodies return to conclude their journey, there are many waiting for just these riches.

Among the first to take their share are the bears. They wade through the shoals, they position themselves in strategic places along rapids, they smell the air. And they wait. They are slim and hungry after a winter's hibernation and a lean spring diet, and they need to fatten up before the next cold season. Some of the salmon they consume on the spot. Others they take away, dining only on choice morsels such as male brains or female roe. What is left of the fish will fall to the ground, where more hungry creatures await. Flies and other insects are drawn to the carcasses by alluring smells of decay, and as they come to eat, they lay their larvae into the flesh. When the larvae hatch, they are born into an all-edible nursery. In heavy rains, the larvae are washed back into the streams, where young salmon and other small fish will prey on those who received their own bodies from the organic inheritance of the salmon elders. A cycle, one of many, has come full circle.

A History of Separation

Beginning with the Greek thinkers Leucippus and his student Democritus in the fifth century BC, our Western metaphysical tradition has said that Being can be broken down into small, indestructible units called atoms. The Greek word for "atom" has the same connotations as the Latin word for "individual": The Greek ἄτομος or *atomos*, can be translated as "indivisible," as can the Latin word *individual*. Both concepts—one physical, the other metaphysical—say that there are fundamental, indivisible units or entities of being. According to Leucippus, if you take a piece of wood and cut it into half, and then you cut the half in half, and so on, you will eventually end up with parts that can no longer be divided. In this way, Leucippusean thought is a kind of *reductionist thinking*—it reduces all of Being to these indivisible units. This thinking has deeply influenced our narrative canon, and it also underlies Descartes's later thought. Callicott suggests that this atomistic understanding of Being—the "ontology of atomistic materialism"—leads directly to the "mechanical solution" of something like Newton's gravitational theory and Descartes's *cogito, ergo sum.*

The modern, atomistic metaphysics of the machine became a powerful, monistic lens through which our modern narrative tradition came to know the more-than-human world, and our place within it. As Callicott writes, this lens continues to affect our worldview: All natural processes, carried out by the "very elaborate machine" of a natural being·such as a plant or animal, "can be exhaustively explained reductively and mechanically." According to this view, the beings that inhabit the Earth have "no essential connection to one another. They are, as it were, loosed upon the landscape, each outfitted with its (literally God-given) Platonic-Aristotelian essence, to interact catch-as-catch-can."[4]

The ecologist and philosopher Paul Shepard similarly writes that our modern worldview "presumes naively that the landscape is a roomlike collection of animated furniture,"[5] into which we embodied creatures are arbitrarily placed like cupboards and sofas and armchairs, dwelling therein but being essentially replaceable. Likewise, the late British philosopher Anthony Quinton of Oxford University has written: "[in the Newtonian understanding, the] world resembles a warehouse of automobile parts. Each item is standard in character, independent of all other items, in its own place, and ordinarily unchanging in its intrinsic nature."[6]

This Newtonian-Cartesian-materialistic-atomic conception finds expression in the proposition that it might be time to let wild salmon go extinct if they get in the way of the industry. It also finds expression in the thinking of companies that strive to redefine salmon as terrestrial creatures. The concurrent alienation of farmed salmon from their ecological communities and the demise of their wild cousins are implied in the narrative structure of an atomistic and materialistic metaphysical monism. Remember the perplexing ontological situation of the AquAdvantage salmon. They are identical insofar as they serve the interest of the company. They are also special insofar as they serve the interest of the company. They are contained elevenfold, with barbed wire defining the horizon of their lives. Their dead will be buried beneath a layer of plastic. They will be flown from Canada to Panama and then back to US markets. Like other landlocked salmon, they are objectified by a logic that adheres to a monistic narrative structure. Likewise, wild salmon are increasingly at odds with this same understanding. They "get in the way of the fish farming industry."[7] The next conceptual step is as logical as it is frightening: "We must be ready to sacrifice wild salmon."[8] Once salmon cease to exist conceptually, it is but a small step until they cease to exist.

When young fish ingest the larvae, they are swimming in what Ned Rozell of the Geophysical Institute of the University of Alaska calls a "salmon tea"—a river infused and saturated with salmon.[9] This tea's ingredients also includes the newer eggs of the next generation, which are rich in protein, fats, and nutrients—compact meals for other stream dwellers. There are also the decaying carcasses of those salmon who have successfully procreated but exhausted themselves in the effort.

None of the Pacific salmon species live past procreation. About one Atlantic salmon in ten will survive and travel back out to sea. Because not all Atlantic salmon die after they have spawned, some nutrients will be carried back to the sea instead of settling into the land. But out of ten spawning Atlantic salmon, nine will remain to nourish the ground. The story of salmon who give their bodies to become the land—this account of mutual giving and taking, of participatory shapeshifting—is a universal story, a story that spans the entire northern hemisphere, from the Bering Strait to California and to Japan, from Maine to Iceland, from Portugal to the Norwegian Arctic.

If you have ever enjoyed a glass of chardonnay from a Californian vineyard, chances are you have delighted in the fruit of grapevines nourished by salmon. Joseph Merz of the East Bay Municipal Utilities District in Lodi, California, and Peter Moyle of the University of California at Davis have studied rivers in California's Central Valley and found precisely that. They compared the salmon-rich Mokolumne River with the salmon-poor Calaveras River. Mokolumne River has consistent Chinook salmon runs even today, although the numbers of returnees have fallen from about a million Chinook salmon to one-fifth of that. It produced wine grapes that received up to one-fourth of their nitrogen from the oceans, brought to these land-bound, immobile plants by the nutrient-rich bodies of spawning salmon. The researchers observed many scavengers feeding on salmon carcasses along the Mokolumne, and they found salmon jawbones scattered in Mokolumne vineyards.[10]

Studies on the Olympic Peninsula on the North American West Coast, where the Elwha River flows, found that scavengers take almost half of the coho salmon carcasses from small streams in the area. A quarter to 90 percent of all the nitrogen in the hair and bones of grizzly bears of the Columbia River basin is of marine origin. Salmon account for more than 90 percent of the nitrogen of Alaskan brown bears.[11]

The nitrogen and the phosphorus brought upstream by salmon can penetrate the riparian soil up to 70 meters from a stream, and in Southeast Alaska, traces of such marine-derived nitrogen have been found in trees and shrubs growing up to half a kilometer away. Up to a third of the nitrogen of valley-bottom forests along salmon streams has been calculated to have migrated upriver in icy, slapping bodies. Trees that grow alongside salmon streams can grow up to three times as fast as those that grow beyond the reach of salmon. The Sitka spruce in Alaska, for example, will take only one hundred years instead of three hundred before becoming a tree big enough to create a pool when it falls into the stream.[12] Such pools will then become the graveled birthplace of new salmon. The salmon not only feed the entire riparian community. They also create, in death, the habitat that their kin will need to thrive, many salmon generations in the future.

Cycles within cycles, like raindrops on a turbulent river.

Relationships Are the Primary Reality

The science of ecology initially arose within the mechanistic ideology. Early ecological thought portrayed the biosphere as what Callicott calls a "mechanical Leviathan"—a gigantic machine consisting of many smaller machines. "Like a planet in its orbit or a gear in its box, each species exists to perform some function in the grand apparatus," wrote the Swedish botanist and zoologist Carl von Linné (1707–1778), clearly showing ecology's historical indebtedness to Cartesian thought.[13]

Ecologists overhauled this ontological canon in the early twentieth century. The American plant ecologist Frederic E. Clements (1874–1945) conceived of another fundamental ontological metaphor for ecological thought: the superorganism. According to Callicott, Clements understood the Earth as "a vast 'comprehensive' organic being," one that "cannot be reduced to the sum of its parts."[14] By theorizing the Earth community as one superorganism, Clements breathed coherence, affinity, and life back into a worldwide array of atomistic, individual spare parts. In Clements's theory, the ancient, pancultural metaphor of the world as a living organism was revived from within the very culture that had abandoned the metaphor three centuries earlier in favor of the "world-machine."

In the century since, the reclaimed metaphor has been further thought through both in its ecological implications and its "ontological overtones,"[15] developing into a theory of complexly flowing and interacting energy fields. In mid-century, the ecologist Aldo Leopold called "land" the "foundation of energy flowing through a circuit,"[16] a circuit composed of living creatures and the soils and waters and air in which they dwell. Not long after, Paul Shepard wrote that Nature "is best describable in terms of events which constitute a field pattern."[17] In the 1970s, the American biophysicist Harold Morowitz expanded on this emerging field theory, using it to directly challenge classical atomistic metaphysics: "The reality of individuals is problematic because they do not exist per se but only as local perturbations in this universal energy flow."[18] Within this emergent ecological understanding, energy flows through the biological system as creatures come into being and die away, lending their bodies to support the lives of others. In this context, sharply delineated, atomistic divisions make no sense.

At the same time, the chemist and engineer James Lovelock was working with microbiologist Lynn Margulis on the articulation of Gaia theory, which broadens Clementsean thought to a planetary level. Gaia theory suggests that the planetary community as a whole enables the continued flourishing of life on this particular planet within this particular solar system, here in this rotating arm of the Milky Way galaxy. The various kingdoms of life are deeply integrated with the so-called abiotic aspects of the Earth—the hydrosphere, the lithosphere, and the atmosphere—through complex feedback relationships. The result of these cycles of participation is larger than the sum of its parts. The living Earth, or Gaia, has been described alternatively as "self-regulating"[19] or "(trans)formative."[20] The Gaian planet seems to have maintained habitable conditions within the biosphere's fragile envelopes over vast stretches of geological time. The biosphere's self-transformative quality also cannot be predicted by knowing any of its constituent parts alone. It is an emergent property. It can be thought of as an ongoing creative or autopoetic act—as Abram has suggested, as an act of "self-birthing."[21] The various regions of the planet—the biota, the atmospheric layers, continental plates, clouds, volcanoes, the oceans—engage one another in such astonishingly creative, dynamic, and integrated patterns of participation that it becomes implausible to uphold the conceptual separation of "living" and "nonliving" Earth domains.

Also in the 1970s, the Norwegian philosopher Arne Næss turned away from his preoccupation with logics, semantics, and Gandhian thought toward a synthesis of his previous work in a truly ecological philosophy. Næss's ecological turn began with an ontological renouncement of classical atomism and the embracing of a field theory. At a conference in Bucharest in 1972, Næss publicly announced his turn, saying: "The deep ecology movement rejects the human-in-environment image in favor of the relational, total-field image: organisms as knots in the biospherical net or field of intrinsic relations. . . . The total-field model dissolves not only the human-in-environment concept, but every compact thing-in-milieu concept."[22]

Though ecological thought was initially indebted to a Newtonian-Cartesian worldview based in classical atomism, it has since emerged as a theoretical alternative that could point toward an alternative narrative. The ecological thinking that emerged in the 1970s suggests that we think of Being not as an assortment of "entities," but as polycentric, richly layered, nonhierarchical, intricately structured, and ultimately unified processes, in which each thing is implicated by all others, and none can exist outside the fluid context of all others. In the embodied thick of living communities, it is implausible to conceptualize discrete units. Atomism is an untenable position.

Who Is She?

As a keystone species, salmon exist within a field of fluid relationships, but this understanding is in tension with the needs of the salmon industry. It is common in the industry to treat salmon as discrete units that can be modified, displaced, and permanently contained. But from an ecological angle, this is troubling. What remains of salmon if they are defined in isolation from the larger context of the biosphere? Where does salmon end and the land begin?

Thomas Berry and Derrick Jensen have both pondered the Being of nonhuman animals in ways that might help us begin to consider the Being of salmon. "The face of the bear, the size of her arms, the structure of her eyes, the thickness of her fur—these are dimensions of her temperate forest community," Berry writes. "The bear herself is meaningless outside this enveloping web of relations."[23] Jensen considers the difference between animals in the wild and animals in zoos, and notes that we develop strange ideas about animals that we see in zoos: "We learn that you can remove an animal from her habitat and still have a creature. We see a sea lion in a concrete pool and believe that we're still seeing a sea lion. But we are not. That is all wrong. . . . A sea lion is her habitat."[24]

How can we begin to think (as well as intuit, feel, and sense) salmon not as discrete ontological units, but as radically relational creatures: intelligent beings who engage creatively, cunningly, and fluidly in the metamorphic depth of the real? How can we alert our awareness more capably to an embodied sentience so unlike our own, one attuned to the Earth's magnetic fields, to the oceans' somatic soundworld, to the vaulted movements of celestial bodies, to drifting weightlessly in their waterworld as if gravity did not exist?

Drivers of the ongoing commodification of the living creature are trying to make a case that salmon are what Jensen describes as "meat and bones in sacks of skin." You can lock up salmon in a steel-reinforced concrete tank lined with rows of barbed wire; you can engineer these units of meat and bones into triploidy to make them sterile; you can rewrite their genetic memory to conform with the grammar of your cultural imperatives, and you would still, according to some in the industry, have a salmon.

But that is not so. A salmon is a sensing, sensitive being making conscious choices inside a sensuous aquatic world she coinhabits with other sentient beings. She is also the unique smell of that one estuary, that place where she once passed from her freshwater youth into her saltwater adulthood. She is the magnetic field that spans from pole to pole and sends waves of recognition through her sensing body on her long journey back home. She is the anticipation of riverside humans who hungrily await her return from the ocean. She is the pace at which riverside Sitka spruce metabolizes icy skin into wooden bark, and the way in which grizzly bear metabolizes her into the hair and fat that will sustain grizzly through another cold season. She is the seasonal swelling of her river's currents. She is her river's topography, its resistance, its moods. She is all that.

To define the creature away from her web of relations, and the web of relations away from the creature, is to open the way to exploiting this creature, to diminishing her, violating her, abusing her, denying her. We are who we are in relation to others. Salmon are who they are in relation to trees, rocks, ravens, and rivers.

Historically, seabirds and migrating fish are estimated to have moved 150 million tons of phosphorus from the oceans to the continents, simply by feeding out in the food-rich open waters and then returning the nutrients to the land through their feces and flesh. These numbers have since collapsed. This seabird and fish pump is now estimated to transfer 96 percent less phosphorus than it did during historical peaks.[25] Phosphorus is one of the building blocks of life, an essential food both for plants and animals. Without it there would be no DNA and no RNA, no bones or teeth, no insect exoskeletons, no energy transfer inside our cells—no metabolism whatsoever.

Seabirds and migrating fish now move only about 5.6 million tons of phosphorus from the oceans to the continental shores every year, as seabird colonies and anadromous or land-going fish populations across the globe are declining.[26] About one in four seabird species is currently classified as threatened. As for the estimated 110 species of anadromous fish that migrate from the oceans to breed in rivers and die there, populations have collapsed to less than 10 percent of their historical numbers both in the Pacific Northwest and in the northeastern as well as the northwestern Atlantic—nearly the entire range of Pacific and Atlantic salmon.[27] Seabirds and anadromous fish connect the vastly diverse ecoregions of the oceans and the land. The fish are especially important, simply because they move so much farther inland than the birds.[28]

The physical causes for the decline of salmon are well documented and complex. Overfishing, ocean acidification, river damming, logging, irrigation, grazing, pollution, urban and industrial development—all have contributed their share. In Washington, Idaho, Oregon, and California, salmon have disappeared entirely from about 40 percent of their historical range. The food that once found its way to bald eagle, grizzly, coyote, gray wolf, wolverine, Pacific giant salamander, Pacific Coast aquatic garter snake, Harlequin duck, Sitka spruce, redwood, Douglas fir, and many others,[29] is now disappearing. On the Olympic Peninsula, home to the Lower Elwha Klallam, at least twenty-two different animal species have been observed to feed on salmon carcasses.[30] The desertification of the oceans is trickling over into the land, one malnourished creature after another.

Ted Gresh and his colleagues write that "[u]ntil recently, the decline of Pacific salmon was primarily treated as an economic or aesthetic loss. Biologists now suspect that salmon depletion is also an enormous ecological loss."[31] This massive privation impacts not only the larger riparian community. It also spells trouble for the salmon themselves: Juvenile salmon require the same marine-origin nutrients that come from the bodies of their dying ancestors.[32] When the spawners never arrive, when the rivers remain vacant, when bodies do not decompose, all will suffer the consequences. Gresh and his colleagues recognize in this "an ecosystem failure that has contributed to the downward spiral of salmonid abundance."[33] It is a silent collapse a hundred years in the making and still under way. It is a systemic change that "may push the possibility of salmon population recovery to self-sustaining levels further beyond reach."[34]

A Story of Interbeing, in Five Acts

In the ecological story of salmon as a keystone species lies a beautifully Gaian tale, a story of interbeing, a story of cooperation and abundance that we have only begun to piece together. Let's gather around some of the evidence, as if around a campfire at night. It would be the appropriate setting for our story: the circle, held together by the fire at its center.

Act 1: Wolves and Moose—Isle Royale, the largest island of a 450-island archipelago in upper Lake Superior, at the border between the United States and Canada, has been studied since 1958 because of the predator-prey relationship between the moose and timber wolf populations that inhabit it. Isle Royale is far enough removed from any other piece of land to prevent significant wolf and moose migration to and from the island, so that the population dynamics are impacted entirely by the interplay between moose (the wolves' near-exclusive food source) and wolves (the moose's only predator). Historically speaking, neither of the animals were indigenous to Isle Royale. The first moose arrived early in the twentieth century after swimming the distance from the Minnesota lakeshore. For the next half century, they had no predators, and they thrived and multiplied. Around the middle of the century, wolves arrived, trotting across an ice bridge that had formed from the Canadian border some 35 kilometers away.

In November 2009, three wildlife biologists from Michigan Technological University published a report discussing the fact that decaying moose carcasses produced hotspots of biodiversity across the island.[35] The authors located the site of each of the 3,600 moose kills over the entire fifty-year period of the project, and they measured nitrogen, phosphorus, and potassium levels in the surrounding soil. They also looked at which microbes and fungi were inhabiting the soil and studied the leaf-tissue of large-leaf aster, a common food of the Isle Royale moose.

Soils in places where carcasses had lain contained between 1 percent and 600 percent more nitrogen, phosphorus, and potassium than soils from surrounding sites. There was also, on average, a 38 percent increase in bacterial and fungal acids, which was a sign that bacteria and fungi were finding the sites more favorable to eat, grow, and multiply. In the plants they studied, the authors found a 25 to 47 percent increase in nitrogen levels as compared to sites where no carcass had been. Plants with higher nitrogen content will in turn attract more hungry moose in the future, and the feces and urine they leave behind will further help plants, fungi, and bacteria to prosper. Another cycle, one of many, had come full circle.

Act 2: Forests, Snow, Albedo—Forests, when they feed on sunlight and carbon dioxide, enrich the atmosphere with oxygen and produce sugars that will feed fungi and animals. They also draw huge quantities of carbon dioxide from the air. Trees store the carbon in their massive and long-lived bodies. When they die, the carbon sinks into soils, peatlands, or aquatic sediments, and remains bound there for a long time. Such forests are known as carbon sinks—huge lungs that draw more carbon from the atmosphere than they breathe back into it. The northern boreal forests are such powerful drivers of this pump that they account for almost one-third of all terrestrial or land-bound carbon.

Moose are the dominant mammalian herbivore of the northern boreal forests, and wolves are their top predators, the ones who keep moose populations in check. Without the wolves, moose populations would grow to several times their former numbers.[36] When too many moose eat too many young trees, forests cannot grow to the same height or develop the same kind of canopy closure that forests with intact moose-wolf dynamics can. This leads to warmer and drier soils, which not only increases the risk of forest fires but also lowers the overall forest's productivity. Studies show that boreal forests with intact wolf-moose populations can absorb an additional 42 to 95 percent of all of Canada's carbon emissions.[37] Fewer wolves equals more moose equals less productive forests equals a possible change in climate. That's how powerful this wolf-moose-forest carbon pump is.

As we train ourselves to take a Gaian view of life, relationships are never clear-cut and linear; surprising feedback loops are the rule. In a letter to *Nature*, climate researcher Richard A. Betts suggested a different kind of feedback loop that could fully offset the boreal forests' entire absorption of carbon. Spruce trees and other evergreens of the taiga have evolved to withstand massive snowfalls and not break under the pressure of the snow. They shake off the heavy snow in a simple and elegant gesture, a bow of the long and flexible branches that frees them of their burden. This survival strategy also affects the planet's climate: Seem from above, boreal forests are relatively darker than the treeless tundra, especially during the months when snow covers the ground. This means that the forest will absorb more sunlight from space and increase regional temperatures, leading to less snow, leading to higher temperatures, setting in motion a cycle of participation that could fully overwhelm the cooling initiated by the growth of the forests. The full entanglements of wolves, moose, forests, and climate remain an open mystery.[38]

Act 3: Whales—Sunlight reaches only the uppermost part of the ocean's water column. This thin upper layer of the ocean is called the euphotic zone, or the zone of "good light." Good for anyone whose life depends on sunlight, which is just about everyone on this planet.[39] The euphotic zone is home to the community of microscopic, photosynthesizing organisms collectively known as phytoplankton (from Greek *phyton* "plant," and *planktos* "drifters" or "wanderers"). Here they drift and drink sunlight and absorb carbon dioxide dissolved in the water, and in turn they produce organic compounds. Phytoplankton is a hugely diverse community of tiny plant beings, composed of five thousand different species, and it is the primary food producer of the oceans, the basis of the entire aquatic food web. It also contributes roughly half of all the oxygen presently available in the atmosphere.

These tiny green drifters cannot live on sunlight and carbon dioxide alone. They also need minerals such as nitrate, phosphate, or iron. That puts them in a peculiar situation. From above, the sunlight confines them to the upper water column, while from below, gravity pulls down unceasingly on the nutrients they depend on. Nitrate, phosphate, and iron are too heavy in the water. Given a chance, they will sink down into the abyss. But without them, creatures who must drink sunlight cannot live.

It gets worse: When fish or zooplankton (that complementary community of microscopic animal drifters) defecate, their feces sink, and with them, more nutrients are lost. When they die, their carcasses also sink, and more nutrients sink out of reach. Some of the phytoplankton even sink while alive, unable to stay afloat in the upper water for very long. This process is known as the "biological pump," and it spells trouble for the phytoplankton. This is where the large ocean mammals come in.

Many of the great whales dive deep into the black abyss to hunt for krill or squid, but they must also repeatedly return to the surface to breathe. When they do, they poop—huge clouds of liquid feces that float on the ocean surface. In fact, they have a habit of pooping at the surface. Like salmon when they migrate inland, the whales also bring rich foods into comparatively poorer geographical ranges—in this case, up to the marginal zone where photosynthesizing, single-celled algae can survive. Their fecal plumes, or clouds of poop, are rich in nitrogen and iron, the stuff that the single-celled algae are craving. As they move up and down the water column they also mix the water, and this vertical mixing helps some of those little drifters stay afloat in the upper column. The technical term for all of this is *trophic cascades*. It means more whales equals more plankton equals more *anything* in the oceans![40]

More plankton also means that more carbon dioxide will be bound in their tiny single-celled bodies. When they die, their bodies will sink to the bottom. This removes carbon from the atmosphere, depositing it in ocean-floor sediments for thousands or even tens of thousands of years. More whales equals more plankton equals more food for everyone, but this equals also a cooler planet! When the numbers of great whales were at their historical peaks, they might have removed several tens of millions of tons of carbon from the atmosphere each year.[41]

Act 4: Algae, Birds, and Clouds—Phytoplankton might be small, but they are not defenseless. Some of them emit a chemical, dimethylsulfide, when they are under attack. This attracts predators who come to hunt the smaller animals that attack the single-celled plants. A number of seabirds can smell dimethylsulfide in the air, among them albatrosses, petrels, shearwaters, and fulmars. They know that they must only follow the scents that waft up their beaks, and they will find their next meal. The giant seabirds and the single-celled algae have a pact that works for both. These birds, too, will take a shit, and this will mean even more food for the plants that attracted the birds in the first place.[42]

James Lovelock, in collaboration with colleagues, described a peculiar process in a report in *Nature* published in 1987: The very gas that attracts seabirds while protecting algae under attack also produces clouds that will ultimately help cool the planet![43] When dimethylsulfide evaporates into the air above the water, it oxidizes to form a sulfate aerosol. These aerosols act as tiny particles that help clouds form just above the ocean surface. This matters, because while oceans are a dark surface, clouds are a far brighter surface. Large areas of cloud cover increase the planet's surface albedo: They reflect more sunlight back into space. Like whales, tiny algae fighting for their lives appear to be able to help cool the planet.

Act 5: Glimpses of Gaia—There is something beautifully subversive about these tales. Each one is a refreshing dip into the astonishing interconnectedness of our small, spherical planet. Single-celled algae seed clouds that reflect sunlight back to space, salmon nourish bears or vultures or vineyards or temperate rain forests, whales bring more fish to the entire ocean and cool the planet, and every act in our story invalidates the ideology of separation; each draws us deeper into a felt kinship with the more-than-human world. This Earth can no longer be thought of as a dead object composed of discrete entities, a big lump of rock that can be conceptualized in hierarchical terms with one thinking ape outside and on top of everything and everyone else. That story has

exhausted its credibility. The world is a network of inseparable relationships, complex, nonlinear, nonhierarchical, and often unexpected. As
Evernden writes, "an individual is not a thing at all, but a sequence of
ways of relating: a panorama of views of the world."[44] Can boreal forests
be conceptualized in the absence of wolves? Can salmon be thought of
as if they existed outside of the deep seas and the rivers? Can climate
dynamics be adequately addressed if we don't also question our treatment of whales and wolves?[45]

The old urge to withdraw from the world remains powerful. Descartes
tried very hard to withdraw. The salmon industry continues to reenact
the old drama, fueled by the epistemological optimism that characterized Cartesian thought. It might almost seem as if the industry is on the
winning side of history: It has the benefit of habit, of economic power,
of political muscle, of technological leverage. It creates a false sense of
infinite abundance, not to mention entitlement and success. This adds up
to a story so commanding, so big, that it almost seems like the real world.
Sadly, it all adds up to tragedy, too. Gaia hasn't evolved to fit that story's
confines. Earth hasn't evolved to host storytelling animals who forever
try to run from it, to deny physical reality, to live outside life's many
beautifully and delicately spun cycles of participation. Relationships are
the primary reality. That is the baseline that the old story cannot erode,
try as it might. Abundance for our own kind is within reach, but only
when we forsake the compulsion to step outside the living world. True
abundance is within reach only inside this Earth.

Our spiraling fireside story not only challenges the atomistic interpretation of salmon in feedlot practices. It also adds substance to our
earlier contention that this phenomenology of story must resist the
objectification of salmon and the partitioning of humans, and the way in
which both these acts of separation have been pursued in the Cartesian
tradition. If we can no longer uphold a sharp ontological distinction
between self and other; if it becomes more prudent to speak of relational
selves; if minds participate in, and are nourished by, the larger web of
relations to the same degree as bodies are; if both minds and bodies are
not discrete entities but aspects, angles, or local perturbations of one
emergent, stratified, differentiated, strangely animate planet—then can

we not assume, with the words of the American educator David Orr, that "it is not possible to unravel the creation without undermining human intelligence as well"? Can we not say that the discovery of the primacy of relationships directly challenges those who suggest that human well-being will not be affected by the disappearance of salmon (or wolves, or whales, or any other fellow creature)? If it is true that we, like salmon, are fully embedded in the living matrix of this small planet, entirely therein and entirely thereof, then can we not say that we are only healthy to the degree that salmon are healthy? Perhaps we are human only to the degree that we give salmon the opportunity to be salmon.

———

Who is she, again? She is the gravel bed where she hatched from her tiny, transparent egg. She is the long succession of moons that wax and wane throughout the Arctic winters. She is the April storms and the August heat, the spring floods and the September rains. She is the whales who fertilize the drifting plants who will be prey to the crustaceans and fish she hunts. She is the algae who excrete the oxygen that fires her flesh. She is the clouds above the surface that help keep Earth cool in the face of an ever-growing sun. She is Earth's expanses of liquid water, she is its surface temperature, she is its drifting continents. She and I are both Earth, thinking and sensing and feeling itself inside each of us.

CHAPTER FIVE

The Sea in Our Veins

The more clearly we can focus our attention on the
wonders and realities of the universe about us, the
less taste we shall have for destruction.

RACHEL CARSON

My daughter Kaia Isolde was born just over three years ago. Her world still seems to be growing a little more spacious every day. Her tiny body is so delicately attuned to me and to her mother that it becomes difficult to say where one ends and the other begins. It has been over three years since I cut the umbilical cord between them. Scissors in hand, I watched it pulsate in small waves. Then, from one moment to the next, it stopped moving. The midwife placed her hand on my shoulder, and I understood. It was time. I cut the cord, and just like that, my daughter was separated, on her own. But somehow, it seems, our three bodies have remained inseparable, deeply entangled and connected. I suppose in some delicate ways, they always will.

My astonishment drifts back in time. The blood that flows through my girl's veins, where are its headwaters? In the womb, drinking and eating and breathing were all one for her; it was what her mother gave her; it was what trickled through that small umbilical river from which everything the child was had flowed. Are these her headwaters? What came before?

If her blood flowed from mother's blood, must I not also trace back the flow of *her* blood? This would take me upstream to her mother's umbilical unity with *her* mother. And so I follow the watershed up through the generations, upstream and past the point where I know their stories, their faces, their names. The bodies begin to metamorphose before my eyes, until the upright walk of distant grandmothers cowers forward and downward, mother by mother by mother, into a four-by-four tread, past all of her mammal grandmothers and even past all of her reptile grandmothers, until limbs reform into the fins they once were, and body parts as separate as breasts, teeth, and hair all grow back into the early skin of ancient mothers from whence each emerged.[1] These mothers inhabit the brackish water of Pangaea's primordial shoreline. They are called *Tiktaalik*. They have all the bones of my girl's upper arm, her forearm, her wrists, and her palms, but they also have scales and fin webbing. They have large, heavy heads and an even larger, powerful tail. They are fish on their way to colonize the land.

Three hundred and eighty million years have since elapsed. Life on Earth has been around nearly ten times as long, but on the scale of her life, that geological time is entirely unfathomable. So many generations have come and gone that I am left with a sense of vertigo. Yet neither my girl nor I have ever left the waters of that first ocean. I need only to pinch my index finger with a needle, squeeze out a drop of blood, and let my tongue absorb my body's taste. I am reminded that we both still inhabit those waters. As Lynn Margulis and her son Dorion Sagan write: "The concentrations of salts in both seawater and blood are, for all practical purpose, identical. The proportions of sodium, potassium, and chloride in our tissues are intriguingly similar to those of the worldwide oceans. . . . No matter how high and dry the mountaintop, no matter how secluded and modern the retreat, we sweat and cry what is basically seawater."[2]

This is not a coincidence, but a living memory. The sea flows through our veins. My little one and I are both creatures of the sea. Blood gives me an immediate taste of our ancestry, but there are other ways in which the ancient waters still flow inside of us, and we inside of them. Margulis and Sagan go on to observe that "fertilization betrays a common aquatic ancestry for every living animal. The essential act of animal creation still always takes place in water. Derived from sea, river, pond, or the body's own tissue fluid, sperm and egg always meet in a wet environment."[3]

Unlike all who now live on land, salmon have never left those waters. Immersed in the lubricants that their eggs and sperm require, salmon continue to make love the way they have always done. Their lovemaking is with the surrounding medium itself, where there is no strict outside and no strict inside, where bodies are at once discrete and porous, where the environing terrain is at once habitat and womb. The female releases her eggs into the redd she has dug into the gravel bed. It is an act of trust that the water will receive her children, protect them, and nourish them with the oxygen they need. When the female releases her eggs, a male discharges his milky cloud into the redd. The water carries the cloud, and it drifts past the couple. By the time it has been carried farther downstream, new lives have been conceived, each a single cell at first, then a cluster of cells, then a body breathing, moving, itching with small life. Once hatched, they will consume the yolk sack that was their mother's farewell gift. After a few weeks they must learn to feed themselves. They will mature, and eventually they will set out on their unthinkable journey into the ocean, where they will grow into formidable adults and learn the skills it takes to navigate thousands of miles of unmapped seascape. Even as their journey takes them far away from their home river, none of them will ever leave the body in which they were conceived. They remain, for life, immersed inside the womb of the water. This is why male salmon have no penises: They simply don't need them. They are already where the penis would take them.

As Margulis and Sagan write, every time one of us is conceived, whether we will grow into wolf, reindeer, human, golden eagle, or muskox, father's sperm and mother's egg meet in water. The land-bound female mammals re-create the ancient ocean in the womb. The feathered and the cold-blooded creatures re-create the ancient ocean in the fragile oval spheres that they lay in nests, tree cavities, burrows, or grasses. Some—like humpback, killer whale, or harbor porpoise—have gone full circle and returned to the sea after a few hundred million years away, carrying a little ocean within while also being carried by a vast ocean without. Whoever we are, each of us begins our life in the ocean. All of us are migrants, wanderers who set out on a long journey but return to our headwaters at the moment of procreation and entrust those who come after us to the ancient sea.

Humans Are Tuned for Relationship

"Humans are tuned for relationship."[4] Thus begins the first book of philosopher and cultural ecologist David Abram. Abram's writing is permeated by the observation that in insisting so stubbornly on separation, we moderns ignore the primordial ways in which we are entirely immersed in the more-than-human world. Abram follows Heidegger's lead in conceiving of the deconstruction of metaphysics as a deconstruction of the mind-world divide. But Abram goes a step further. In the context of worldwide ecocide, Abram suggests that transcending received dualisms must mean adopting a sense of wholeness that embraces the entirety of the planetary community. At the heart of his phenomenological project is the careful and diligent work of reimagining the human fully within the context of the larger living planet. Abram writes: "We are human only in contact, and conviviality, with what is not human."[5]

Abram's fascination is with the ecology of perception, or the ecological dimensions of sensory experience. He is interested in how our eyes, our skin, or our ears bind our nervous system into the depth of the encompassing ecosystem. His approach often centers on a deep fascination with the ecology of language—how what we say has a profound influence on what we see, hear, taste, or even think. There are ways of speaking, he suggests, that can open and enhance the interchange between our senses and the body of the sensuous world—its plants, its other animals, the air, the soil, the oceans. But there are also ways of speaking that close down our senses and that blind us to anything that is not human or of human design, ways of speaking that deafen our ears to any voice that does not speak in words.

Unlike Heidegger, Abram's foremost interest is not in Descartes, but in a young Portuguese Jew born in the city of Amsterdam fourteen years after Descartes dreamt in the city of Ulm and five years before the *Discourse on the Method* appeared in print. Named at birth Benedito ("The Blessed") de Espinosa, this Portuguese-Dutch boy is now commonly known as Baruch Spinoza, lens grinder and philosopher. Spinoza saw himself indebted to the thinking of his French predecessor, Descartes. Like him, he harbored a deep affinity for rationality. Spinoza biographer

Steven Nadler writes that Spinoza strongly believed that "one should strive to understand God or Nature, with the kind of adequate or clear and distinct intellectual knowledge that reveals Nature's most important truths and shows how everything depends essentially and existentially on higher natural causes. The key to discovering and experiencing God, for Spinoza, is philosophy and science."[6] Although methodologically indebted to Descartes, Spinoza pursued a different agenda. As Abram describes it, Spinoza did not agree with Descartes's dualism and suggested "instead that mind and matter were . . . simply two different attributes, or aspects, of one and the same substance, which he called *Deus, sive Natura*, 'God, or Nature.'" Abram goes on to observe that "*every* material body or thing, for Spinoza, had its mental aspect—all things were ensouled."[7]

Abram points out that Spinoza's fundamental identification of God with Nature "earned him an abundance of scorn" not only in his local community but "throughout Christian Europe, ensuring that very little of his work could be published during his lifetime."[8] When he was twenty-three years old—the same age as Descartes when he pondered *Quod vitae sectabor iter?* (What path in life shall I follow?)—the Jewish community issued a *cherem* against Spinoza, a full-blown ban from all communal life, expressed in raucous language.[9]

Spinoza became an influential philosopher regardless. Hegel famously claimed that "You are either a Spinozist or not a philosopher at all,"[10] and Abram writes that "some of the most important theorists of our own era claim Spinoza as their progenitor and inspiration."[11] Yet Spinoza has not become paradigmatic like Descartes. At the beginning of the twenty-first century, modernity still inhabits a predominantly Cartesian cosmology.

It seems that there is no "objective" reason why Descartes, rather than Spinoza, became paradigmatic for philosophy, for science, and for culture of the past four hundred years. Descartes was not, objectively speaking, more "right" than Spinoza. In fact, modern ecological thought would suggest that he was actually wrong. But philosophers are received from within the narrative structure of their time. Abram suggests that it was not so much the separation of mind and body, but the more profound separation of mind from the material world that "ensured the rapid ascendancy of the Cartesian worldview and its long success in the West." This separation allowed the rational mind of the Enlightenment "to pursue its giddy dream of comprehending, and mastering, every aspect of the material cosmos. Descartes' segregation of the mind from the body, in other words, was but a means to a grander end; it authorized the modern mind to reflect upon the material world as though it were not a part of that world—to look upon nature from a cool, detached position ostensibly outside of that nature."[12] It might be that Descartes became canonical because his work resonated more immediately with intuitions his contemporaries had already harbored. It might be that Descartes's merit was that he structured the nuanced perception that was already being felt—the profound distrust in the life of the senses. He might have given his followers the impression, in Neil Evernden's words, that they finally knew what they had only suspected before.[13]

But an alternative storyline was struggling to be born there at the dawn of modernity. Abram writes that "[Spinoza] alone saw that the human mind could never be reconciled with the human body unless intelligence was recognized as an attribute of nature in its entirety."[14] Such was Spinoza's understanding of wholeness. He offered it at a time when Europe's brightest thinkers were keenly aware that an earlier sense of wholeness had been lost. This alternative storyline was stalled. Yet to this day it lingers on in the shadow of that other story—humanity-as-separation—offering itself as a choice worth (re)considering.

As Heidegger tells us, the "phenomena" of phenomenology do not show themselves. They withdraw into concealment and linger there, in plain view yet enigmatically out of sight. This is also the nature of being in narrative. Being in narrative, we do not easily read narratives as narratives. The narratives are hidden in plain sight. Working in the phenomenological tradition after Heidegger, Abram seeks to circumvent the difficulty of being in a phenomenon by coaxing that phenomenon out of hiding in a particular way: He attempts to see the ordinary in its extraordinariness.

This is how we can begin to clear space for the revival of an alternative narrative. If Cartesian epistemology appears so ordinary as to become invisible, then we must strive for a subtle shift of perception, so that what appears ordinary (and thus intangible, obscure) might instead reveal itself in a different fashion: as odd, idiosyncratic, anything *but* ordinary or natural. This is what we must do with the story of separation. Pressure it, remain skeptical to it, defamiliarize ourselves with it, refuse to accept it as normal.

The Real in Its Wonder

Neil Evernden observes that "the response of both [Romantics and phenomenologists] to the threat of 'custom' or the 'natural attitude' is to counsel *wonder.*" "*Wonder,*" in Evernden's view, "*is the absence of inter-pretation.*"[15] Earlier, I noted the classical Greek roots of this counsel to wonder. Heidegger also recognizes its central importance. In *Being and Time*, he first writes of "the contemplation that wonders at being, *thaumázein,*"[16] but his interest in that topic did not emerge again until a lecture series held ten years later.[17] In that 1937/8 series, he called *thaumázein* the "'basic disposition' [. . . that] transports [thinking] into the beginning of genuine thinking."[18] Heidegger follows Aristotle here, suggesting that philosophy emerges from wonder. "Wonder," he writes, "is the disposition in which and for which the Being of being reveals itself. Wonder is the tuning within which the Greek philosophers were granted the correspondence to the Being of being."[19] Philosophy finds expression "in the manner of correspondence, which attunes itself to the voice of the Being of being."[20]

What Heidegger calls attuning to the voice of the Being of being, Abram translates into his interest in language. He suggests that received linguistic concepts isolate us when we are drawn so close to them that we lose our critical distance. His work seeks to "find fresh ways to elucidate . . . earthly phenomena, forms of articulation that free the things from their conceptual straitjackets, enabling them to stretch their limbs and begin to breathe."[21] The intent is to suspend ideas, to step before interpretation, and to nurture a sense of perplexity at that which commonly reveals itself as ordinary, thereby withdrawing from our critical gaze.

The final chapter of Abram's *Becoming Animal* has the telling title "The real in its wonder." It is an intricate dance between discursive and excursive writing, between philosophical reflection and descriptions of an excursion through a North American landscape. Having described his journey through an immense, five-hundred-thousand-year-old claystone formation at the seam between the American Midwest and the High Plains, Abram reflects on "the weird security of realizing that one is part of something so damned huge."[22] This "weird security" can be read as a response to that early modern sense of utter insecurity that Descartes once felt. Realizing that "one is part of something so damned huge" has the opposite effect to trying to step outside of and rise above the world. To experience oneself as "a clutch of sodden soil and hollow bones," to observe how the winds that animate the nearby treetops also animate the inner landscape of one's own mind—this can have a profoundly liberating effect. We need not use Heidegger's martial language and think of ourselves as "thrown" into the world. We can also observe how we are "held and sustained by powers far larger"—the gravitational embrace of Earth's larger body[23] or the air that sustains our blood and thought with small, successive portions of solar energy, metabolized for us lungbreathers by the photosynthesizing folks in the land and water. Such palpable phenomena can potentially draw us back inside, inside what Abram often speaks of as "the larger body of the Earth." Such phenomena can give us a heightened sense of relationship and of security.

The motivation to rethink Earth not as a *res extensa* but rather as an inside, as a kind of body, suffuses all of Abram's writings. He observes at length the fluid and participatory entanglements of our mindful bodies with the living terrain. The eye for detail is programmatic, for his phenomenology originates in concrete, lived encounters with the world. Evernden's suggestion that "wonder is the absence of interpretation" aptly describes this phenomenological practice. Abram strives for that absence of interpretation by carefully crafting a style that startles the senses, plays with convention, and keeps the reader alert. His writing invites us to leave well-trodden thought-paths, to step away from metaphors with a long trajectory of philosophical interpretation.

Leaving a well-trodden path to strike out into the wilderness is not without risks. Evernden observes that "in comparison with other academic specialists [the phenomenologists'] work may . . . seem poetic."[24]

This is a common objection, and one that I, too, am familiar with from my work. The style of phenomenologists is a widespread cause of misunderstanding, and it is received with a mixture of hesitation and frustration. There is still a certain expectation that "good" academic writing should be less poetic and entertain a cool detachment. The poetic is associated not with the clarity of thinking commonly ascribed to academic rigor, but with "subjectivity" or fuzzy thinking.

Yet, if we hope to explore the possibility of other narratives—the "range of alternative 'stories' we could have adopted as our own"—we must find a suitable way of looking at and relating to the world. As Heidegger indicates, phenomenology strives to attune itself to the voice of the being of Being. It asks to be granted the correspondence to the being of Being. The required disposition for this is *thaumázein*—wonder. As Evernden observes, "For us wonder is a harbinger of hope, since it reminds us of our ability to suspend belief."[25] Through wonder, we are given the possibility of renewing philosophy and choosing a new story by deconstructing our received metaphysics and suspending our belief in the dominant narrative.

The nearness to poetry is not a coincidence; it is a program. Heidegger described the primordial significance of the poetic as that through which "everything first steps into the open."[26] The poetic is a way of speaking that enables us to give expression to the primordial experience of being within a world that is laden with meaning, a world that refuses to be reduced to objective data, where hiddenness and knowledge are not opposites but complementary aspects of being in the world. It is through poetic saying that we are able to "bring forth" that world, to bear witness to its inherent self-emergence. That bringing forth happens by tuning in. The poetic does not create resonance as much as it resonates with being. It does not make as much as it receives. It is not a flight of fancy. It is the fluid, ongoing, dynamic attempt to serve that receptivity.

Heidegger also points out that the poetic is not an exclusive faculty of poets; it is rather a certain bearing, or attitude, toward the world. Anyone might choose to encounter the world poetically. Anyone might nurture a deep sensitivity for Earth's immanent autopoiesis, its poetic upsurge into being. As the American writer, critic, and naturalist

Joseph Wood Krutch points out, this can be a difficult task: "It is not easy to live in that continuous awareness of things that is alone truly living. . . . the faculty of wonder tires easily. . . . Really to see something once or twice a week is almost invariably to have to try to make oneself a poet."[27] Those who write in the service of such a poetic attitude might become exceptional instructors in mindful participation with the environing terrain and all those who dwell within. The gift of such poets as Terry Tempest Williams, Stephanie Mills, Renée Askins, Derrick Jensen, Mary Oliver, or Jeannette C. Armstrong is not to describe the world but to reveal it in its meaningfulness. The offering they make to the cultural imagination is precisely their readiness to bear witness to the meaningfulness that emerges spontaneously out of participating in a living community that is far greater, more diverse, more complex than the community of humans. Their craft bears witness to the world's immanent, self-birthing wonder.[28]

For the reader, this implies a certain willingness to also suspend belief, a readiness to not simply dismiss the writing as "nothing but poetry." Evernden formulates this poignantly: "A prerequisite to wonder is loss of complacency. Before anything can be experienced as wonderful, one must resist the temptation to dismiss it as nothing but something else, that is, as something 'un-wonderful.'"[29] The attuning that phenomenology strives for is a participatory process not only between the phenomenologist and the world, but between the phenomenologist, the world, and the reader. It is a happening that unfolds in between. The poetic quality of phenomenology might not be a problem with the writing, as much as it is a problem with the reader's expectations.

Such a phenomenological gaze inevitably contains an improvisational element. To be in relationship means to be a participant, and that means to speak and to be spoken to in return. It means to let go of the notion that by stepping into objectivity, we can control all aspects of the inquiry. When we stop trying to control, Abram suggests, we see that in observing we are also being observed. When we speak to the world and act within it, the world will respond. Our task as writers becomes to practice the subtle craft of respectful, mutual participation with other-than-human expressive styles, or eloquences, whom we encounter. There cannot be any final certainty, and the only thing we can be certain of, after all these centuries of living Descartes's dream, is "the world disclosed to us by our direct sensory experience."[30]

The critique of humanity-as-separation permeates Abram's work, and the phenomenological method becomes his tool for deconstructing that alienating expression of humanism. The scope of the problem is reflected by the scope of critical means. Rational argumentation is one way of using language. But it is not the only way of speaking, nor is it always the most appropriate way of speaking. If the deconstruction of separation—the reweaving of reciprocity—is to succeed, it must be mirrored in both speech and methodology.[31] Different ways of speaking and of knowing might be more accurate under different circumstances. Not all ways of knowing might be relevant at all times. This does not mean that one should simply add a poetic element so as to make the writing "more beautiful." As the Caribbean-American writer and civil rights activist Audrey Lorde has said, "Poetry is not a luxury."[32] It is never adornment. If the poetic element is used, then it is only because there might be some aspects of the inquiry that cannot be spoken of more clearly. It is always motivated by the wish to know more fully, in more subtle ways.

CHAPTER SIX

Being Human

Erase my words and see:
Fog is moving across the meadows.

RAINER MARIA RILKE

C onsider for a moment a peculiar ancient Indo-European word: *dghem*. Lewis Thomas calls the word a "linguistic fossil, buried in the ancient root from which we take our species' name."[1] It has since metamorphosed into such contemporary terms as *Earth* (English), *erde* (German), or *jord* (Norwegian). It has also transmuted into other terms such as *humus, humanity, humanist,* and *humble.* Thomas observes that in this very brief list lies "the outline of a philological parable."[2] In the very name we give our species lies dormant the primordial experience of a deeply reciprocal participation between ourselves and the ground beneath our feet—between the human animal and the Earth.[3]

In her book *The Winter Vault*, the Canadian poet Anne Michaels succinctly recaptures this ancient intuition:

How much of this Earth is flesh?
    This is not meant metaphorically. How many humans have been "committed to earth"? From when do we begin to count the dead—from the emergence of *Homo erectus*, or *Homo habilis*, or *Homo sapiens*? From the earliest graves we are certain of, the elaborate grave in Sangir or the resting place of Mungo Man in New South Wales, interred forty thousand years ago? An answer requires anthropologists, paleopathologists, paleontologists, biologists, epidemiologists, geographers . . . How many were

the early populations and when exactly began the generations? Shall we begin to estimate from before the last ice age—though there is very little human record—or shall we begin to estimate with Cro-Magnon man, a period from which we have inherited a wealth of archaeological evidence but of course no statistical data. Or, for the sake of statistical "certainty" alone, shall we begin to count the dead from about two centuries ago, when the first census records were kept?

Posed as a question, the problem is too elusive; perhaps it must remain a statement: how much of this Earth is flesh.[4]

This intuition has left fossil traces in the sediment beneath our linguistic topsoil, traces of a deeply reciprocal engagement between our mindful animal bodies and the microbial, fungal, plant, and animal communities that make up the soil beneath our feet. Others have sought to express that felt reciprocity in speech. In her piece "Sleeping in the Forest," the Pulitzer laureate Mary Oliver ponders this enigmatic and never fully fathomable metamorphosis between humus and human, between Earth and earthling:

I thought the Earth remembered me,
she took me back so tenderly,
arranging her dark skirts, her pockets
full of lichens and seeds.
I slept as never before, a stone on the riverbed,
Nothing between me and the white fire of the stars
but my thoughts, and they floated light as moths
among the branches of the perfect trees.
All night I heard the small kingdoms
Breathing around me, the insects,
and the birds who do their work in the darkness.
All night I rose and fell, as if in water,
Grappling with a luminous doom. By morning
I had vanished at least a dozen times
into something better.[5]

A century and a half ago, Walt Whitman sounded his "barbaric yawp over the roofs of the world," pondering the same theme:

> *A child said,* What is the grass? *fetching it to me with full hands;*
> *How could I answer the child? I do not know what it is, any*
>   *more than he . . .*
> *And now it seems to me the beautiful uncut hair of graves . . .*
> *What do you think has become of the young and old men?*
> *And what do you think has become of the women and children?*
> *They are alive and well somewhere;*
> *The smallest sprouts show there is really no death;*
> *And if ever there was, it led forward life, and does not wait at the*
>   *end to arrest it;*
> *And ceas'd the moment life appeared.*
> *All goes onward and outward—nothing collapses;*
> *And to die is different from what any one supposed, and luckier . . .*
> *The last scud of the day holds back for me,*
> *It flings my likeness after the rest, and true as any, on the shadow'd wilds;*
> *It coaxes me to the vapor and the dusk.*
> *I depart as air—I shake my white locks at the runaway sun;*
> *I effuse my flesh in eddies, and drift it in lacy jags.*
> *I bequeath myself to the dirt to grow from the grass I love;*
> *If you want me again look for me under your boot-soles.*[6]

The British reverend John Donne spoke the following during a sermon in March 1619, only months before Descartes would be visited by his dreams:

> Take a flat Map, a Globe *in plano,* and here is east, and there is west, as far asunder as two points can be put: but reduce this flat Map to roundnesse, which is the true form, and then east and west touch one another, and are all one: So consider mans life aright, to be a Circle, *Pulvis es, & in pulverem reverteris, Dust thou art, and to dust thou must return; Nudus egressus, Nudus revertar, Naked I came, and naked I must go;* In this, the circle, the two points

meet, the womb and the grave are but one point, they make but one station, there is but a step from that to this.[7]

Some eight years earlier, William Shakespeare offered a comical variation on the same theme in his play *The Tempest*. The invisible spirit Ariel sings these words to Ferdinand, prince of Naples, who believes that his father has drowned:

Full fathom five thy father lies;
Of his bones are coral made;
Those are pearls, that were his eyes:
Nothing of him that doth fade,
But doth suffer a sea-change
Into something rich and strange.
Sea-nymphs hourly ring his knell:
Hark! now I hear them,—ding-dong, bell.[8]

At the genesis of Christian cosmology, there are also the names of the first two humans to walk the Earth. The Hebrew word *adama* means "earth" or "soil." *Eve* is a translation of Hebrew *hava*, or "life." As the geomorphologist David Montgomery observes: "[T]he union of the soil and life linguistically frames the biblical story of creation."[9]

———————

The philological parable unfolded in these citations—call it "Of Manure and Men"—tells of an intimate relationship between Earth and earthlings, between humus and humans. It hints at an original, pancultural experience, a primordial insight into the human situation. Some might say: This is nothing but poetry! It is intangible, and flowery. It has no import for the real world. It cannot withstand the hard-gained ground of pragmatic decision making.

Recent studies suggest the opposite. In 2007, Montgomery published his book *Dirt: The Erosion of Civilizations*, in which he concludes that the current way in which modern humans encounter soil around the globe is "suicidal."[10] "Considered globally, we are slowly running out of dirt," he writes. "Unless more immediate disasters do us in, how we address the twin problems of soil degradation and accelerated erosion will eventually determine the fate of modern civilization."[11] He does not linger with the resonance of that desperate word, "suicidal," long enough to catch its metaphysical overtones. The immense erosion not only implies a sweeping loss of food, water, burning material, and animal and plant companions—it also implies a gradual erosion of ourselves, a loss of our humanity. This is the truth that the poetic, pancultural insight points to: Earth and earthlings are entirely inseparable from one another. The essence of one is determined by the quality of its relationship to the other. We are Earth. Earth is us.

In the real world of business-as-usual, soil, like salmon, is overwhelmingly considered as "commodity," "inanimate matter," or "resource." Such is the objective and precise language of the realist. Even those with the best intentions—those who wish to put their skills in service of the larger community of life on Earth—are often drawn into the gravitational sphere of that perceived realism, abandoning what sensuous kinship they once might have experienced with a gushing mountain stream, a flock of migrating geese, the awe-inspiring authority of an ageless glacier, or the fugitive presence of a snowflake on their outstretched palm. They speak instead of the living world as if it were mere inanimate stuff; they speak of "sustainable resource management" or "maximum sustainable yields."

It turns out that the poetic mode is not a luxury. The realist might be the actual dreamer, and poetic thought might in fact be a pragmatic, down-to-earth, humane, and practical inclination. Patrick Curry, speaking of Abram's work, writes that "in contrast to [such ugly, obfuscating, subjective words as] 'ecosystem services' and the like . . . what Abram calls 'the practice of wonder,' being fully embodied and embedded, is actually as solid as it gets."[12]

Evernden called wonder "a harbinger of hope."[13] The poetic disposition might not be unlike the seed banks of Svalbard and elsewhere where we seek to protect our heirloom seeds from calamity. There, like here, we place our hope in the bosom of the Earth. There, like here, we tend to our seeds carefully, sending them out into landscapes in need of rejuvenation and diversification.

Or perhaps it would be more accurate to speak of the poetic disposition not as seeds, but as manure. Like manure laid on the ground, the poetic mode nourishes other discursive spheres—such as "degrowth," "Earth jurisprudence," "transition towns," or "rewilding." It vitalizes and emboldens the work of skilled experts from a variety of disciplines. Just as manure replenishes the soil with essential elements, making it richer, more diverse, and more resilient, the poetic disposition replenishes the human situation with primordial experience, feeding our humanity, making us richer, more diverse, more resilient—and more realistic.

———

In December of 1816, the young English poet John Keats wrote "the poetry of Earth is never dead." He was right, of course. The poetry of Earth has never been silenced. Even from within a monocentric narrative tradition with its anthropocentric "realism," the poetry of Earth, of soil, still rises forth. The primordial insight of the human-Earth connection does not belong exclusively to the distant past. It is an insight into the Earth's inherent self-emergence into the presence. The world always emerges forth into being right here and now in a continuously creative process of becoming. This is how each human generation originally encounters Earth: in its self-emergence.

Being human, we receive the nourishment of thought, insight, and awareness primordially from the Earth itself. What originally sustains us is the poetry of the Earth, even as the beleaguered Earth struggles hard in our time to never be dead. The poetic mode is precisely that which enables us not to starve and not to lose sanity. It allows us, as Heidegger says, to tune in to that voice of the being of Being, to tune in to a cacophony of Earth voices, as Earth articulates itself through dialects as different as the waggle dance of honeybees, the seasonally composed song of male humpback, human speech, or the vitalizing resurgence of rains after a period of drought.

A chapter on dirt, on soil, on humus might seem rather landlocked for a book on salmon, but it might be the best way to ask an essential question: What could it mean to rethink the question of knowledge from the ground? Earth itself might be speaking through us. Perhaps barren landscapes out there both result from and cause an equally barren landscape within. To better understand what it means to be human, we might have to start by understanding humus. And to know salmon more fully, we need to ask again what it means to be human.

Listening

The Cartesian formula *cogito, ergo sum* is also a form of poetry; it, too, is a way of letting Earth step out into the open, though in a rather peculiar way. Through it, Earth reveals itself exclusively through the constricted eloquence of rational human thought. But to the degree that rational thought reveres itself as the sole locus of ingenuity—to the degree that it strips the Earth of its poetic disposition—Earth becomes revealed primordially as absence. Earth's inherent self-emergence becomes concealed. The poetry of Earth falls silent.

In response to this, we could conceive of the phenomenological project as an ongoing attempt of re-membering—"put together again, reverse the dismembering of."[14] The intention is to reverse the dismembering of human voices from the great chorale of Earth voices, to step outside the Cartesian-mechanistic narrative, and to step inside what Abram calls this "unfathomable upsurge of existence itself."[15]

The question of rethinking the question of knowledge from the ground now assumes a slightly altered shape: How can we best lend words to this unfathomable existence, this voice of the being of Being, as it surges up through the porous texture of the soil, through the porous skin of our bodies, through the sensuous and synesthetic engagement of our minds with the living land, and, at last, into human speech? How can we best lend words to this upsurge of existence, this bubbling well of Being, which might be more accurately called *a well of becoming*? The task is to be attentive to the felt texture of that upsurge as it voices itself through human speech. The task is to aid Being into unconcealment through the eloquence of human speech, to honor our human expressiveness while also trying to eavesdrop on the many other eloquences with which the Earth speaks itself.

Speaking in this manner begins in listening, and its primary task is to speak without hampering our chances to continue listening. "Listening" is a shorthand for the practice of sustained attentiveness that keeps our ways of speaking and thinking porous and permeable, so that they might be nourished with the primordial expressiveness of other Earth eloquences. We listen to the odd barrenness of a snowed-over forest lake in mid-January, steeped in impenetrable morning fog, an austerity that leaves our thoughts vague and disengaged as we ski quietly and quickly. We listen again to the profusely flavored air by that same lake on a sunny afternoon in late March, just as the ground is being liberated from the clutches of frost, and wetness, rich earthen scents, and arousal are spilling from the ground and gushing into our agitated flesh, a sensuous spring tide that overflows the formerly frozen landscape of our imagination, flooding our awareness with sensations that, too, had been held underground, inaccessible to us during the long months of darkness and frost.

If Earth is the primordial wellspring of thought and of speech, then must we not allow our speech, every now and then, to slip back into Earth? If the landscapes of our imagination are resilient and diverse to the degree that we let them be nourished by the soil's original eloquence as it surges up from beneath our feet, then must we not let go of words, of concepts we might hold dear, so that they might be folded back in, broken down, digested, fermented, composted, rejuvenated? Perhaps we must let them shapeshift from their particularly human form into that complementary, more-than-human form—Earth? Therein they might nourish the ground, which in time will conspire with us to birth new words, new thoughts, new concepts that are fresh, vibrant, meaningful. What if we considered human speech and human thought not to be our exclusive "possession" but rather our contributions to larger gift cycles, larger cycles of participation?

———

In the final footnote of *The Spell of the Sensuous* lies a skeleton key, carefully placed there by Abram to help unlock the book in its entirety:[16] "In contrast to a long-standing tendency of Western social science," he writes, the book does not attempt "to provide a rational explanation of animistic beliefs and practices. On the contrary; it [presents] *an animistic or participatory account of rationality*."[17] Animism, or the notion that *everything speaks*, is a "wider and more inclusive term"[18] than rationality, preceding it, underlying it. Reciprocal and participatory modes of experience "still underlie, and support, all our literate and technological modes of reflection. *When reflection's rootedness in such bodily, participatory modes of experience is entirely unacknowledged or unconscious, reflective reason becomes dysfunctional, unintentionally destroying the corporeal, sensuous world that sustains us*."[19]

Abram uses the words *animistic* and *participatory* interchangeably. Both share the understanding that anything can engage our mindful bodies in a fully reciprocal exchange, a broken tree stump no less than a mountain meadow of heather, a nightingale no less than a wood stack, an overcast sky no less than a moonless night saturated in starlight. Anything can summon our gaze, claim our attention, keep our senses captive, draw us near, repulse us, dull us down, invigorate us, distract us, sharpen our attention. Anything is fully and entirely animated. Perception is not a unidirectional and static gaze by a "subject" onto mute and passive "objects." It is in no way bound by a monocentric theory of knowledge. It is a fully participatory exchange between a complexly layered multiplicity of animate powers, each unique in its expressive style and pace, each partaking deeply and fully of a common mystery, illuminating some local region of that mystery. The very notion of static objects—objects resting dead and immobile next to one another like abandoned car wrecks in a junkyard, or like furniture in a roomlike landscape—becomes implausible. Each thing is always already in motion. Each thing is imbued with agency. Through that motion, pulse, or animation, each thing has the power to engage us.

Some things, like lightning or a breaching killer whale, surge into our awareness so promptly and forcefully that they leave us speechless. Other things, like the echolocation of bats, pulse so rapidly that they try the outermost reaches of our perception.[20] Others, like radio waves, are faster still, pulsing at intervals so brief that we can no longer participate in their upsurge with unmediated senses.[21] Then there are those that—let us try that more animated pronoun: *who*—beckon to us so leisurely, so unhurriedly, that they might seem inert from our vantage point. The discursive style of trees is so unlike that of us two-legged apes—so very slow, nonverbal, and different—that it is exceedingly difficult to hear them speak at all. Socrates notoriously said to his good friend Phaedrus that although he was a lover of learning, trees and open country teach him nothing, whereas men in the town do. Yet trees speak, each in their own while, each of their own storied and seasoned participation with the land, addressing our imagination as we seek their nearness for shade, fruit, mushrooms, the sustenance of their wooden flesh, or simply for company. A sandstone cliff is slower still, protruding from the Earth's vegetative cover like a giant scar in the landscape. Yet that rock formation also moves in the firm manner of its kin, speaking in its own oddly stiff tongue. But sandstone cliffs perform rap songs in comparison with the Gregorian chant of continental plates, drifting across the molten interior or our planet, animate powers who are so vastly unlike ourselves that we cannot encounter them in their lethargic upsurge with unmediated senses.

When they do speak, suddenly, in voices we apprehend corporeally— when their inert tension is released as earthquake or as tsunami, or when the molten interior of the Earth erupts as volcano—they speak with voices so commanding, so powerful, that we struggle to apprehend them with human speech. In Isaiah 30:33, volcanoes are compared to the "breath of the Lord." Similarly, the Great Lisbon Earthquake of 1755 became known as the "wrath of God." In both cases we see the acknowledgment of an animate agent (or power, or voice) unlike our own, larger and older, a voice who could easily crush us, of such potency that only the mythical imagination might be spacious enough to gauge it.

We human apes are nested firmly in the midst of a vast multitude of overlapping pulses, rhythms, beats, sighs, beeps, rumbles, waves, moans, hiccups, burps, buzzes, and hymns, witnessing the never fully fathomable upsurge of existence—what Keats called the poetry of Earth—from the vantage point of our mindful bodies. We are bounded by a perceptual horizon that we do not easily stray beyond unless we engage in the art of storymaking to imaginatively wander beyond that horizon, or in the equally delicate art of toolmaking to momentarily stretch our perceptual participation and eavesdrop on articulate powers beyond our bodily, sensuous attunement to the land's voices.

Given the ways in which the legacy of the Cartesian tradition tends rather to conceal the Earth, to reveal it only as absence, I wonder: How can the vast and vastly diverse upsurge of existence voice itself to moderns *other than as silence*? How can a truly animate Earth, with a soil and a sky and an ocean that speak, make itself heard if moderns are so busy with their own rational minds—developing, bulldozing, blasting, paving, renovating? How can waterfowl, insects, boreal forests, or coral reefs continue to follow the trajectory of their own surge into existence if moderns allow nothing to grow old?[22] What becomes of this multiplicity, this living diversity, when a single mode of upsurge seeks to monopolize all others? The irreversible loss of biodiversity we are witnessing today appears to be directly connected to one creature's stark insistence on the singular importance of its own, rational attunement to the world. But how can a truly modern people think that rational thought, cut off from its source, will not eventually wither alongside all those others who are now withering or who have already slipped away?[23]

Inside the Depth of the Upsurge of Being

I stand in the middle of the forest clearing outside the house. My eyes are drawn this way and that, lingering on the compost heap across the lawn before being drawn toward the woodpile that lines the western wall of the barn. Since I stacked those logs last fall, the pile seems to have gone awkwardly crooked. The ground must have shifted as frost first spread like a cold breath across this forest clearing early last winter, settling again a few months later, leaving all that had been resting on it in slight disarray, recomposed. While I wonder at this strangely slow yet perceptible malleability of the solid Earth on which I stand, my eyes are drawn away, off toward the northern edge of the clearing, down the slope, past the enormous boulder formation that protrudes from the ground, past the tallest tree, the spruce who has not only grown remarkably large but also branched out into nearly two dozen daughter trees (all of them still connected to the maternal elder through long, thin wooden limbs, former branches that now have morphed into oddly wooden umbilical cords, visible above the ground).

At last, my eyes come to rest on the pond. This pond marks the northern boundary of my place. I notice a hesitant unrest in my feet; they appear to be off to somewhere, but are unsure which way to walk. I notice that not only my eyes are drawn toward the pond but also my ears. Only now do I notice the quacking of ducks, somewhere on the far shoreline. The pond's pull becomes more insistent, and I notice how my thoughts, too, have begun wandering toward it. Just a few days ago it had still seemed strangely inert and silent, muted by a lingering sheet of late-winter ice. It had not seized my attention as it does now. Then I notice that my feet have set off in motion! This is odd, for I had not been aware of any conscious decision to start moving. By the time my understanding has caught up, my feet have fallen into an easy stride. Ah, so that is where I am headed—I am on my way to the water!

As I approach the pond from a distance, I notice how its surface stirs with agitation. The ice lingers only in tree shadows. Tiny, deep-blue waves lap, stirred by a cold midday wind that stirs me, too. The air collides, ever so gently, with my cheeks, rippling across the curved surface of my face in a succession of minute shivers, sending cold chills down my back. I hear the wind pass through the spiraling interior of my ear; I sense it sink further in until it spreads, like a cold breath, bringing me into resonance with it. I feel the urge to take a deep breath. More ripples, spreading through me. Firing my blood, setting my imagination ablaze. I stand here at the edge of the pond, this spring day, and I am strangely excited.

This is one of the first days of sun after weeks of continuous cloud cover. Water vapor had hung so thick and heavy above the land that the narrow space between the clouds above my head and the ground below my feet had seemed to contract, to diminish. There was no felt progression of the sun in the sky, or the moon, no felt texture to time. And just as the land lay in this state of thick idleness, so, too, the ways of the community of two-leggeds seemed to stiffen and grow wearier. Conversations became fewer and less frequent; eyes met less often. Not that I noticed. The change was so slow that it never materialized into a conscious thought or a choice. I was inside the changing mood of the land, unable to distance myself from it.

Now the sky has opened up. Incessant sunshine is flooding the land, saturating every nook and corner of my vision with color and movement, invigorating my ears, infusing me with a new sense of curiosity, agility, and restlessness. I ache to see, to hear, to touch, to smell. I also want to *be* touched.

Time's texture is more stratified and more interesting now than just a few days earlier. Two coal tits carry twigs, moss, and feathers into the birdhouse outside my library window. I am content to stand for a while and follow their busy doings. I imagine what enchantment will soon seize them as spring advances, rousing their flesh, stirring their affection, urging forth new creation.

Long, earthen lines crisscross the lawn like scars, as they do every spring after the snow has melted. They tell me where mice must have scurried back and forth through their tunnels for the past months, safe and warm beneath the insulating blanket of snow. The first ruby red tips of rhubarb stretch upward, carrying my thoughts forward to summer's first harvest, and backward to last summer's rhubarb cakes, rhubarb drinks, rhubarb jams, and rhubarb ice creams. I look up into the bright blue above and see in the radiant expanse a vague shape, round-bellied, hazy. The moon is only two or three days from being full, which reminds me that the equinox is just over a week away, which sets my mind in motion, calculating until I understand why Easter will come so late this spring.

The temporal texture of this moment is one of rich, thick, fluid, participatory expansiveness. As I am drawn deeper into this place, I settle into a larger and more integrated sense of being present. I sense my body with a heightened alertness; I am more curious and alert now than I was just a few days earlier, more eager to encounter those whom I might come across. But the here and now also become fluidly continuous with my expectations of what is yet to come, and with memories of what has been. This, too, is part of my being present, this day by the pond. I notice all this with relief and gratitude. It really has been a long spell of grayness.

I am walking again, and more memories rise from the land. This clover meadow is where the mother deer mourned her fawn's death one

June day, from sunup till sundown, fighting the ravens who had come to pinch a morsel of flesh with which to feed their raven chicks. That lone birch tree over there is where I stood late one moonless August night and saw the red fox come sprinting toward me. As I wondered how long it would take the fox to see me in the darkness, a deer broke through the undergrowth, chasing the fox. They both ran so hard that I could hear their wheezing and panting as they dashed toward me and then past me, an arm's length away, so close that I could have touched their sweating fur. And wasn't it by that stand of trees over there that I startled the bull moose one black November night? I took him so much by surprise that he *exploded* into the thick undergrowth, tearing down the steep slope toward the river, launching himself headlong into the water, landing in a splash amplified by the dark and quiet autumn night.

Still I am walking, and the smell of the path calls to mind earlier spring days like this one, voices, games, colors, bits of conversations with friends, tail ends of songs, someone erupting in laughter, a late-spring snowstorm, and I look away, along the winding path to the south, and once again I see the venomous adder who lay there sunbathing one torrid July day, and I see the stray cat I befriended, sitting there on that same path. I spoke quietly with him as he sat at a secure distance, sniffing the air until he was confident enough to come closer, and closer still, and in the end he was so close that he rubbed his tail against my bare legs. Later he moved in with me, and when winter came he moved back out again, leaving a long, straight trail of cat paws in the fresh snow.

As the path winds across the hilltop, my memories roam, become vaguer as the horizon of the hilltop obstructs my view, limiting the scope of my mind, and my thoughts unravel at last in the distance that I cannot immediately see. It seems relevant that Abram has described Earth as the primary mnemonic or memory trigger.[24] These thoughts and memories, they are inside of me somehow, but it also seems that they are equally inside the land. It seems the land holds on to them even as I wander off, gifting them back to my awareness when I return. But when I receive them back, they are not the same; they are not stern, objective "bits of information." The conceptual separation of a neat inside from a neat outside, of self and other, and in fact of subject and object seems too facile to observe the phenomenon of memory, which I experience as the participation between animate agents. The land and I converse, we negotiate the terms of our participation, and the ground releases memories as I witness the seasons pass. The land in the wake of my presence is a land nourished with memories. And I in turn am richer for having radiated into it.

We two-leggeds of the modern narrative tradition do not easily acknowledge that in nourishing the ground with our presence, with our stories, that same ground will feed our intelligence in return. We do not acknowledge this reciprocity, and we do not acknowledge our roots for what they are; we dismiss them easily as sentimental by-products. We shrug them off even while our embodied minds are trying to grow roots into the local land, into river, into forest, into salmon and moose and salamander, into the mood of this valley, into the distinct undulation of the local horizon.

Even though we might be struggling to speak accurately about the way these roots sprout from our embodied minds, through our skin and into the soil, our bodies will continue to rejuvenate our ancient acquaintance with the land. Our intelligence will go on probing the terrain, patiently searching for those pockets of nourishment on which it once fed, and as soon as the tips of our roots tap that familiar soil, we will taste sudden familiarity. We will be strangely stirred when we set foot again in the town where we grew up, when we breathe in that air, unlike any other. The bend in the road where a family member was killed continues to germinate cascades of conflicting emotions and remains a hotbed of that crushing sense of loss. The harbor where we awaited the arrival of a friend remains a source of excitement and joy. Our lives inscribe themselves into the land through which we have wandered, and the land takes us in, becoming porous to us, the way it becomes porous to the fumbling explorations of tree roots, and it stays around as the locus of our memory, as the earthen and wooden and aqueous atmosphere of our mindful flesh. Or perhaps we can say: The land becomes the mind's tangible terrain, the concrete topography in which thinking and feeling emerge. Perhaps this is what mind always already strives to establish: an entanglement with landscape, a poetics with locality. Perhaps the infinite possibilities of an active, upwelling presence serve as a sort of blueprint for our own experience of being gifted with infinitely potent and creative minds.

We must understand this Earth, this soil, to be fully entangled with what we commonly call time. The neat separation into space and time becomes artificial, and not quite accurate, when seen from within the thick of the presence. The same place speaks with astonishingly different voices under different circumstances. Winter in a place gathers a certain arrangement of moods and recollections around me, spring will secrete another, and summer and fall will emit others still. When it rains, different memories emanate from the land than in summer drought; the electric tension in the air before a late-summer thunderstorm makes my imagination pulsate differently than a snowstorm in March. In each case, my "inner" landscape grows forth fluidly from the larger "inside" of this land itself. Instead of speaking of the received conceptual dualism of a pure "inner" and pure "outer," it is far more parsimonious to say that I participate in a wider, more richly layered, and more enigmatic interiority.

When Ethical Monism Unravels

Every ethical theory sets up a horizon of reciprocal care, which is also known as a moral circle. That horizon or circle is determined by reciprocal participation with other subjects. It is common in ethics to speak of these subjects as "moral agents," implying that agency is an important feature when we define those with whom we actively negotiate our moral imagination.[25] But as the modern mind withdrew into itself, to step out of the thicket of lived experience and into the cooler, less composite, and more austere space of rationality, ontological and epistemological monocentrism influenced the moral imagination, creating an ethical position that we call anthropocentrism. From within an anthropocentric understanding, soil, trees, the air, the waters, or any of the plants or other animals therein were no longer recognized as active interlocutors. The world we coinhabited with blue whales, wolves, buffalo, passenger pigeons, with the other great apes, with cougars, black bears, and elephants, with the great temperate rain forest and with coral reefs, was redefined, in Thomas Berry's words, as a "collection of objects," and not a "communion of subjects."[26]

A certain trend in ecological ethics has been to try to overcome anthropocentrism by expanding the so-called moral circle. The intention is to extend the sphere of care outward, in concentric ripples that reach out to other-than-human creatures who are also known to be able to suffer ("pathocentrism"), to all individual living creatures ("biocentrism"), to larger-than-individual organic wholes ("ecocentrism"), and at last to the biota or even the biosphere as a whole (various types of "holism"). Such work has diversified and enriched the moral imagination. But it overlooks the fact that concentric expansion outward remains inherently engrossed in an anthropocentric—or, we could say—*ratio*centric structure. And this structure is already problematic. Ecological ethics thus conceived does not sufficiently problematize the original epistemological and ontological error of anthropocentrism, namely that the human mind is considered to be the only center, the last remaining stronghold of interiority. While the metaphor of the expanding moral circle works to destabilize the anthropocentric imagination, it might not be subversive enough.

The unraveling of ontological and epistemological monocentrism implies a similar unraveling of ethical monocentrism and the reweaving of a more differentiated, layered, and concrete multicentric ethical imagination.[27] A multicentric ethical imagination, even when it explicitly seeks to break with anthropocentrism, needs to also critique the ratiocentric bias of the Cartesian split. The anthropocentric imagination has been tightly entangled with a denial of those aspects of the human that have been perceived as more animal-like. Anthropocentrism, both as an ethical theory and as cultural practice, doesn't just marginalize other-than-human life. Its humanism is hateful even of our own bodies.

The ratiocentric bias is suspicious of the ways in which, even within the human, different angles of the world flow together, different ways of knowing that color our notion of reality. Whether it is the senses, intuition, feeling, or thinking, each adds a particular hue or radiance to our experience of the real, contributing its local erudition, its capacity for enriching our sentience, overlapping with the others to form a stratified, complex, and synesthetic illumination of the real.

Critiquing the Cartesian split goes hand in hand with a renewed curiosity for those other weird and marvelous animal, plant, fungi, and microbial beings who also experience this world from their own synesthetic vantage points. A renewed curiosity for our own senses, feelings, and intuitions directly inspires a renewed sense of wonder for who else might be there. What is all this boisterous living that is happening right there, right now, just beyond ratio's grasp? What is it like for Earth to experience its richly stratified planetary presence from an other-than-human perspective? Each creature experiences Earth's larger terrestrial body from a unique angle, lending their peculiar sensuous capacities, offering their corporeal attunement with the enveloping terrain, being illuminated about the real in ways unique, inimitable.

What is it like, say, for the living Earth to experience itself as a five-hundred-year-old Douglas fir in the temperate rain forest of the Pacific Northwest? The forest ecologist Suzanne Simard has recently described the way Douglas firs, like other trees, actively intervene in the fate of the larger community of trees around them through enormous fungal networks that connect their roots to other trees, including those of other species.[28] Through these networks, trees are able to communicate, sending messages back and forth, exchanging carbon, water, and nutrients, depending on who is in greater need at any given time. Simard indicates that the trees are actively trying to help each other survive, even across species. She reserves a particularly telling metaphor for the tree elders: "Mother trees" she calls the ancient ones, both for their more-than-average fungal networking with other trees, and also for the way in which, when they die, they send some of their remaining nutrients into saplings just springing up from the ground.[29] As soon as we shift our attention to the possibility of such diversity, of such *agency*, the world seems to enlarge and differentiate. It seems to come alive.

How does each of us earthlings, human and other-than-human, organize experience in the unbroken field of connections that is this planet's biosphere? How can the human animal come to know the world truthfully from within its necessarily local and finite vantage point? How can we speak precisely of our immersion in the larger body of the Earth in ways that not only appeal to the thinking mind, but also recognize the other ways in which our sentient bodies become irradiated from within? What becomes of our notions of truth when we recognize this categorical locality and finitude of the human experience, and when we recognize the lives of trees, fungi, birds, dragonflies, or fish?

All these questions begin by approaching the enfolding terrain ethically, rather than trying to broaden reciprocal care toward a so-called external "environment" from within a monocentric structure. In this perspective, "value" bubbles forth from the complex multitude of the biosphere itself as its thin membrane wraps itself all around the curved belly of the planet. While intrahuman ethics are embedded within this gurgling multitude, and while they remain important for negotiating some aspects of our moral lives, they are nowhere near the single center of the spherical matrix. Such a single center simply does not exist. Given the ineffable multiplicity of *embodied interiorities* on Earth, anthropocentrism is indefensible. As Curry writes: "The only way to resist and ultimately replace the inherently anti-ecological logic of monism is through pluralism. And that means a moral as well as epistemological pluralism."[30]

Affirming Our Membership

The timely contribution of phenomenology to ecological thought is that it nudges the ethical imagination toward the primacy of our perceptual embeddedness as embodied creatures in the living land. It trains us to pay heed to the multiplicity of styles, to other modes of perception and ways of being in the world, to the many angles from which the world experiences itself. It does not imply that humans are anything less than unique or awe-inspiring. In fact, they are even more unique and awe-inspiring than what the ratiocentric bias makes us believe. But this does not make them stand apart, nor does it elevate them. It integrates them more deeply into Earth. It helps the human animal become receptive to the uniquenesses of other marvelous kinfolks, such as mountain trout, or muskoxen, or moose.

Whenever we eat or drink, whenever we pee or defecate, we engage in deeply reciprocal exchanges with other earthlings. When we metabolize proteins out of the amino acids we have received from eating the flesh of other plant and animal creatures, we excrete urea or uric acid, which is a combination of carbon dioxide and ammonia, which in turn is a nutrient consisting of hydrogen and nitrogen. James Lovelock has observed that this is anything but an obvious process, for urea excretion means a significant loss of both water and energy. Why have we and other mammals evolved to excrete nitrogen in the first place? Wouldn't it make more sense, he asks, if our bodies broke down the urea into carbon dioxide, water, and nitrogen gas?[31] Excreting nitrogen might seem like waste, or a loss. But if we did not excrete nitrogen into the soil, says Lovelock, the plants would starve for want of it.[32] In eating, as in secreting, we affirm our membership with the biosphere, from within its midst.

We can say as much about what happens when we die. As soon as we are dead, our bodies begin digesting themselves. At first, internal chemicals and enzymes consume our bodily tissues and organs. Then the vast and diverse community of microbes who have been living on our skin, inside our mouths, inside our nose, and throughout our digestive canal start breaking down our carbohydrates, proteins, and fats. As our bodies consume themselves from the inside, they release acids and gases that produce a foul odor to our own noses, but that odor attracts more creatures, such as insects and flies. Summoned by the bright olfactory signaling of our bodies, scores of these creatures feed on us and lay their eggs into cavities of our muscles or bones, entrusting their future to our rotting flesh. Our bodies are not unlike the Douglas fir elders: In the final passage of our lives, our bodies offer themselves as food to others. They break down the remaining nutrients, making themselves more easily available to others. Parts of our bodies feed themselves back into the soil community of humus, other parts will evaporate into the air to summon feeders, others will be washed away with the flow of water. In dying, we feed ourselves back into the larger flesh of the planet. We gift ourselves back into the wider life of the biosphere. Our disintegrating bodies, like the dying trees, commit themselves utterly and unreservedly to the continuity of the larger terrestrial community of life. One last time, each of us affirms our membership within it.

There is, however, a sense in which we humans remain somewhat aloof, somewhat remote from the busy doings of the other creatures who participate more immediately with soil. Wild boar plows the richly fragrant mud with her muscled and perceptive nose. Cautious wolf follows scent ribbons of earlier passersby through a hazel grove. Ants set up a supply chain between their stately hill-city and the sweet, rotting windfalls scattered beneath a nearby apple tree. Flies buzz around a pile of bear shit in frenzied agitation. Ghostly lynx sits pensively by the scent mark of another lynx, quietly pondering who passed by, and when, and on what errand. All of them are drawn toward, called upon, and summoned by the ground. All of them think their way into the ground; all of them witness thought and ideas rise directly from that ground and into their mindful bodies through taste buds, through the nose, through the eyes, through skin, through memory shaped by, and attuned to, perception. As Abram writes, other animals think with the whole of their bodies.[33]

This style of cogitation is not easy for the human mammal to grasp. Walking upright on our two hind legs, with our heads a meter and a half or more above the ground (a height to which far fewer of the ground's fragrances rise), we are not so easily engaged sensually by the Earth. Abram observes that we human mammals long ago learned to balance on our hind legs, freeing our forepaws to manipulate objects, specializing "in a kind of curiosity that looks at things from different directions, turning them this way and that—a highly visual pondering that seems to unfold somewhere behind our eyes." This, he writes, encouraged "a propensity for detachment that . . . has been greatly intensified in the current epoch,"[34] most recently further amplified by a complex array of technologies that demand of us only "a very rarefied form of intelligence, manipulating abstract symbols while our muscled body is mostly inert." Thinking appears "to have little bearing on our carnal life; it often seems entirely independent of our body and our bodily relationship to the biosphere."[35] Thinking gathers around the central venue that is our lofty head. We do not easily recognize the ongoing, improvised corporeal rapport with the terrain around us, that conversation of our thoughtful flesh with the larger flesh within which we amble.

Yet the bottoms of our feet—those down-turned hands, as Abram calls them in *Becoming Animal*—are among the most sensitive and responsive regions of our bodies. Our feet not only help us balance; they not only give us the freedom to run down a mountain slope, or to stand still among the other upright bodies of nearby trees, such that we become invisible in the landscape—our feet are also in a constant, vigilant, and intimate conversation with the ground. Why else would they be so sensitive? Like other erogenous zones of our bodies, our feet are locations of heightened participation, engaging us not in an erotic dance with a lover, but in an erotic dance with the Earth itself. As we roam barefoot through a cold, dewy meadow, or as we stroll along the ocean's foamy shoreline, or as our naked feet make their way up a steep, rocky ledge, step by watchful step, this particular region of our sentient bodies is constantly alert and receptive—listening, feeling, sensing its way into the wider terrain, *locating us within*. Even if our heads up above in the sky are inclined to forget the richly layered ways in which we are fully of the Earth, our feet continue to affirm our membership. Every time the weight of our bodies comes to rest, momentarily, on the heel, midfoot, and toes of one of our feet, there is a moment in which we are touched by Earth, and in which we touch Earth in return. We walk, skin to sensuous skin.

Defecation, urine, the passage that is decay, our sensuous feet: All are forms of heightened participation. They draw us outward while receiving inward that which is beyond ourselves. They dissolve facile notions of separation. They fine-tune our intelligence as we make sensual, bodily contact with the flesh of others and the flesh of the world. They are manifestations of Earth becoming aware of itself through us—weirdly, marvelously, exuberantly.

Thought that wanders too far from the wellspring of our mindful participation with the land will gradually deteriorate. It will begin to project upon the living land the hardened images of an imagination further and further removed from the tangible ground below our feet. As it does, the ground itself will become depleted and hardened in a tragic and deadly feedback loop. Lester R. Brown, founder of the Worldwatch Institute and founder and president of the Earth Policy Institute, has studied the paving of the planet with asphalt. His documentation of the remarkable amount of land that has been paved illustrates what happens when rational thought loses touch with the Earth: "The United States," he writes, "with its 214 million motor vehicles, has paved 6.3 million kilometers (3.9 million miles) of roads, enough to circle the Earth at the equator 157 times. In addition to roads, cars require parking space. Imagine a parking lot for 214 million cars and trucks. If that is too difficult, try visualizing a parking lot for 1,000 cars and then imagine what 214,000 of these would look like."[36]

This is what happens when rational thought retreats into its epistemological and ethical fortress. The monocentric epistemological imagination and the monocentric ethical imagination are both the result of free-roaming thought believing that it is the central and primary dimension of the world. "I think, therefore I am" is a denial of the primacy of perception, which is the primacy of the body's spontaneous and improvisational participation with the animate Earth. As a result, the modern mind believes that it is possible to fully remake the world in accordance with abstract ideas. Asphalt poured into the once-breathing pores of the ground is an abstract idea concretely articulated. But when the remarkable freedom of our capacity for abstract thought is not coupled with a sense of our ethical obligation to the land, and with a sense of wonder for all those other beings going about their lives, thought can quickly become hugely destructive.

Turning toward the Ground

What could it mean, to rethink the question of knowledge from the ground? To gather the senses that were so thoroughly scattered in the wake of the Cartesian split? As the thinking mind is released from its disembodied exile inside the fortress of *res cogitans*, and as it is absorbed back into the depth of our corporeal participation with the wider animate Earth, our attention is drawn firmly toward the tangible ground below our feet. To walk upon the animate land, to let the voices therein speak, to respond. And to develop a sensitivity for who else might be speaking there. The work of gathering the senses goes hand in hand with a heightened impulse to celebrate and regulate our links to other-than-human beings.

To turn our attention to the ground is a deeply ethical gesture. At the beginning of this chapter I quoted Lewis Thomas, who listed a number of modern words derived from that Indo-European fossil, *dghem*. *Human* was one such word. *Humus* was another one. But Thomas also listed this: *humble*. The etymological heritage hidden in the word "human" suggests that humility in relation to the living Earth is fundamental to being human.[37] When we act as we do in the modern mode, as if we humans alone have consciousness, or as if we alone have language, then in fact we are not being fully human. The very gesture commonly associated with humility is taking a bow. Arching the lofty locus of the thinking mind toward the land beneath our feet, which gives the mind a steady stream of nourishment. Taking a bow: Turning toward the ground. Moving closer to Earth. Becoming, in this way, more fully human.

CHAPTER SEVEN

This Animate Waterworld

Earth, isn't this what you want? To arise in us, invisible?
Is it not your dream, to enter us so wholly
there's nothing left outside us to see?
What, if not transformation,
is your deepest purpose?

RAINER MARIA RILKE

*W*e are 45,000 kilometers above or outside Earth, on a December morning in 1971. Five hours earlier, around midnight, Apollo 17 had launched from Kennedy Space Center. Now the spaceship, still moving at supersonic speed, was leaving Earth's orbit and aligning with an elliptic course toward Earth's ancient wayfarer, the moon. The three astronauts unbuckled and took off their overalls, and they experienced for the first time what it was like to be outside the reach of Earth's gravitational pull. Only now, as their bodies became utterly weightless inside their tiny life-sustaining capsule, had they really left Earth's embrace. They drifted toward a window, like feathers on a draft, and they looked out. What they saw made them ecstatic. Crew member Evans reported, "Yes, the moon is there." And then: "The Earth is . . . that is the Earth! Wow, what a beauty." A few minutes later, commander Eugene Cernan said he could see Earth more fully than she had ever been seen before. "And do you know what . . . she is all alone out there."[1]

At 45,000 kilometers, Cernan, Evans, and Schmitt had flown high enough to see the entire spherical body of the planet through their window, yet they were still close enough to make out many detailed features in the landmasses. They saw the utterly thin, turquoise veneer of the atmosphere as it wrapped itself around the vast luminous globe, more delicate and sheer than a mosquito breath passing across elephant skin. At that moment, Apollo 17 happened to be intersecting the imaginary line between Earth and Sun, so the globe below was illuminated in its entirety, with no eastern or western Earth shadow dimming its full-bellied roundness. There before their eyes lay their entire sphere of existence at a single glance: the Indian Ocean and the Southern Atlantic. Antarctica. All of Africa, cradle of our kind. The Arabian Peninsula and the Middle East, cradle of the agricultural revolutions that set humans on a new path after the last ice age. Whirling bands of clouds, a full-blown cyclone in the Indian Ocean, hinting at the enormous oceanic and atmospheric currents, and at the planet's rotation around its own axis, ever eastbound, sunbound. Nowhere did Cernan, Evans, and Schmitt spot any trace of human life. They saw a fluid, dynamic participation between land and oceans in which water and air appeared to be the single-most decisive agents. The air looked like a great conveyor belt, the medium by which water traveled landward. And there, on land, life appeared to be where water is, and water, where life is. All this against the backdrop of a vast, black emptiness.

The gaze outward into infinite space has long been known to be deeply disconcerting. But here was something else, something that took these three men completely by surprise: Within the black immensity they found something of utter beauty, something that appeared unique: Earth.

The crew took a photograph that day. After NASA published the photograph on December 23 that same year, Blue Marble quickly became the most publicized image in media history, an icon of our time.[2] The image is at once stunning and uncanny, making evident the delicacy of Earth's elliptic journey around its star. It showed that there are definite bounds to this world we inhabit. Earth is an utterly round body. It is also not a particularly large planet. The star around which our planet orbits has a volume 1,300,000 times larger. Any average sunspot roughly equals the size of our planet. We are that small.

The understanding that we inhabit a round world had long been commonplace by the 1970s. It had long ago replaced astronomers' earlier notions of a flat Earth-disc. Even Aristotle had noted the round shadow that Earth cast upon its moon during lunar eclipses and concluded that the only shape that casts a round shadow on a distant surface is a spherical body. Yet it is no exaggeration to say that Blue Marble wholly upturned our understanding of Earth, booming through our perception like a series of mighty earthquakes. For the first time, we saw with our own eyes that the atmosphere is but a fantastically delicate smear on the surface of the globe. We saw the deep blackness around, an infinitude of otherness that harbors no realistic promise of refuge out there. After two and a half millennia of having thought about Earth's roundness (with interruptions, of course, and with fierce debate), it was only now that we fully knew this roundness—not with our rational intellect, but with the entirety of our mindful bodies. And it was just this full-bodied knowledge that enabled us Earth dwellers to grasp what this roundness implies. Fragility, writes historian of culture Ulrich Grober, became a key notion in contemporary interpretations of the Earth image.

The image of Blue Marble enabled us all to intuit the experience of hovering above this planet and beholding it from outside. It let us reexperience, in our own bodies, the sensations that had consistently seized astronauts and cosmonauts as they drifted weightlessly by their tiny spacecraft windows and looked back upon Earth. Take, for example, Edgar Mitchell, astronaut on the Apollo 14 mission: "Suddenly, from behind the rise of the moon, in long, slow-motion moments of immense majesty, there emerges a sparkling blue and white jewel, a light, delicate sky-blue sphere laced with snowy swirling veils of white, rising gradually like a small pearl in a thick sea of black mystery. It takes more than a moment to fully realize this is Earth . . . home."[3] Or here is cosmonaut Boris Wolynow: "When you behold the sun, the stars, and our planet . . . you develop a more affectionate relationship with all that is alive."[4] Astronaut Walter Shierra recalled his multiple space travels in later life: "I left Earth three times . . . and found no other place to go. Please take care of Spaceship Earth."[5] Sometime after landing safely again on Earth, Apollo 17 commander Eugene Cernan said: "We set out to explore the moon. What we discovered instead was the Earth."[6]

Humans have ventured deep into that sea of black, they have gone to a place outside Earth, but the gaze of these astronauts and cosmonauts has consistently been drawn back, earthward. The singular beauty they beheld with their eyes incited their compassion and their care. They saw a fragile pearl in a sea of black. That pearl was utterly round. It was also very blue. Earth, a delicate sphere, a waterworld.

The Animist and the Ocean

To recognize that all things have their unique pace or rhythm or patterned emergence—their speech—is to recognize the agency of all things. The animistic experience undermines moral or epistemological monism and speaks to a pluralism of ways of knowing, as well as a pluralism of the moral imagination. The experience of being human inside the thick of the living community remains centered in ourselves, and we continue to engage with the world morally from within that center. But we also begin to realize the great multitudes of other centers beyond our own perceptual horizon, other loci of experience that are equally of this world. The animistic experience gives credence to the thought that each of us, whether human or salmon, shrimp, bluefin tuna, sea anemone, single-celled algae, or humpback, are Earth experiencing itself from another angle.

What would that be—for Earth to experience its richly stratified planetary presence from any of those other embodied minds who dwell within it? What would it be like to experience an earthen, bodily mindfulness of a very different sort here inside the biosphere? What would it be like to be gifted with another set of sensuous capacities, to hone a different style of corporeal attunement with the enveloping terrain? How do other-than-human earthlings organize experience within this unbroken field of connections?

Such questions begin by approaching every aspect of the enfolding terrain ethically, rather than broadening reciprocal care toward a so-called external "environment" from within a monocentric structure. The animistic experience leaves no space for a so-called environment. If all things are animate, including the oceans, including continental plates, including the long-term hydrological cycle, then there is no "backdrop," no passive background against which things unfold. All things enter the stage! Each thing lends its eloquence, its cadence, its pulse, its line to the great epos that is Earth's storied evolutionary trajectory. Earth, the living planet, becomes a "communion of subjects" rather than a "collection of objects."

Here I return to the ocean. With the eyes (and ears, and skin, and tongue) of an animist I observe and describe how Earth's animation is not restricted to the land, but sweeps vigorously through that other great region of the planet, the oceans. If the oceans and those who dwell therein are also animate, then they cannot be a mere passive backdrop to human-centered concerns. The story of humanity-as-separation continues to have very concrete physical ramifications on the oceans. Overfishing, plastics, toxic runoff, noise pollution, ocean acidification— each indicates that the narrative of separation still treats the oceans as a vast and passive environment or *res extensa*—at times a dump site, at times a treasure trove to be exploited at will. They are there for the rational intellect to study, manage, harvest, consume, trash, ignore. They have no life of their own.

The focus on the oceans calls forth a valuable metaphor: the horizon. In phenomenological terms, the metaphor of the horizon lets us encounter the outer edge of our perceptual experience as a semipermeable seam, rather than a closed-off bounding circle. The horizon brings porousness to our being-in-the-world. Others speak, and it is there at that spongy seam of perception that our embodied mindfulness might encounter other, very different mindful bodies, and participate in reciprocal exchanges.

This will bring us right to the brink of asking: What might it be like, to witness another embodied eloquence dawn at the far end of that horizon? What would it be like to be a salmon?[7]

Waterworld

Once again my gaze hovers outside of Earth. I strain my neck and look up, as if she were hovering above my spacecraft window, and I were below. What I see above me is a huge and round expanse of blue. Water!

So much of it that it gives the globe an azure hue. I imagine this vast blue expanse of water flowing continuously around the planet, its colossal masses passing between liquid and gaseous and frozen phases: as ocean, vapor, steam, cloud, fog, rain, hail, snow; as glacier, lake, fjord, river, cascade, waterfall; as animal body, plant body, fungi, microorganism; and again and again: as ocean. I look up and I see how the entire blueness continuously falls back upon the Earth body. Like the atmosphere, and like the multitudes who inhabit the living sphere's many regions, all the planet's water is held to Earth by that mysterious force of mutual attraction—gravity. Gravity holds on to the oceans, and to the clouds that drift through the air. Gravity lets the rain fall back upon the surface rather than away from it. What if Earth's gravity were too weak to hold on to that water? The planet would become desiccated, a dry lump of rock. And without water, no life.

But the situation with Earth's water might be more precarious than it initially appears. Lynn Margulis and Stephan Harding have observed that "any chemical or physical process that liberates hydrogen from water molecules in principle might lead to water loss from a planet. Hydrogen (H_2) gas has a mass so light that it reaches escape velocity from the Earth's gravitational field."[8] The gravitational field around Earth is too weak to draw hydrogen back in: Isolated hydrogen does not fall back onto Earth, nor upon any of the inner planets. Within two thousand million years, write Harding and Margulis, any of the inner rock planets of our solar system, including Venus, Mars, and Earth, would have lost so much water as to become essentially desiccated. The atmosphere on such a barren planet would be saturated in carbon dioxide, and surface temperatures would be determined primarily by the ever-brightening sun.

This is precisely what happened to both of our neighbors. Venus, Mars, and Earth were all bombarded during their planetary infancy with meteorites, comets, and asteroids that carried massive amounts of water. But Venus, being closer to the sun and receiving more solar luminosity, likely was cast into an early "runaway greenhouse"[9] effect that evaporated huge amounts of liquid water, and once the vapor had risen into Venus's stratosphere, the water molecules were dissociated by ultraviolet radiation. Massive amounts of hydrogen would have escaped into outer space, eventually leaving Venus the bare rock it is today. Mars, too, is likely to have lost its ocean via photodissociation.

Earth alone has managed to hold on to its water, despite two major abiotic processes that hasten the loss of this vital gas on our planet. Both Venus and Mars are subject to the first, which is the photodissociation we just saw. The other is particular to Earth. Deep down on the ocean floors, ferrous oxide in the seafloor basalt reacts with water and carbon dioxide, in turn liberating hydrogen. These two processes combined ought to have desiccated this planet long ago. Yet Earth has remained a wet planet for three thousand million years.[10] In their Gaian interpretation of Earth's "water anomaly,"[11] Harding and Margulis write that it is life itself that has actively worked to retain water on the planet, metabolizing liberated hydrogen via a variety of different pathways, binding the evasive gas back into heavier chemical compounds that have prevented it from escaping into space. On Harding and Margulis's tentative list of metabolic pathways for hydrogen capture, we see that bacteria participate in the bulk of such processes, which is not surprising when we consider that multicellular life only appeared on Earth about a thousand million years ago, after 2.6 billion years of single-cellular life. The metabolic pathways for hydrogen capture are complex and not yet sufficiently understood, but even with the current partial understanding it is safe to conclude that life has drastically slowed down the loss of water throughout its 3.6-billion-year-long evolutionary trajectory. Life itself has kept Earth wet and alive.

Harding and Margulis's tale would be incomplete without another important process—life's active involvement in regulating Earth's surface temperature. Life has not only held on to hydrogen, it has also actively maintained conditions suitable for liquid water. Two factors conspire to render the continuous presence of liquid water throughout geological time unlikely. First, the sun's energy output has been steadily increasing since life first emerged in the early oceans. Our planet now receives about a fourth more heat energy than it once did, for as the sun has aged, it has grown larger and hotter. Combined with the steady output in greenhouse gases through Earth's volcanic activity, this ought to have caused Earth's surface to grow ever hotter, evaporating and losing ever more hydrogen.

We are aware, thanks to the pioneering work of James Lovelock, that our planet has evolved creative responses to this increased heat stress. Gaia has learned throughout this vast span of deep time to maintain her surface temperatures at conditions that are favorable for life.[12] The key insight of Gaia theory is that life not only adapts passively to the inanimate conditions of a so-called environment, but that the biosphere is tightly coupled with the atmosphere, the lithosphere, and the oceans, creating (rather than merely adapting to) a livable planet. This is where the recent pulsation of ice ages comes in. Harding, discussing Lovelock's interpretation of Gaia's temperature regulation, writes that Gaia appears to be in crisis: "[S]he is experiencing some kind of pathology. . . . Gaia's recent wobbles in and out of ice ages may be a clear sign that she is struggling to keep cool under a bright sun—that she is overstretched to the point of instability, with the glacial periods being her preferred state, and the interglacials her fevers during which she hovers dangerously close to catastrophic climatic breakdown."[13]

If, on the other hand, not enough greenhouse gases had been spewed into the atmosphere by volcanic activity, Earth's atmosphere would have cooled, freezing over larger stretches of the oceans, increasing the surface albedo, which in turn would have cooled Earth even further, until, at last, the entire planet had been cast into a permanent frozen state—a permanent Snowball Earth, a frozen white marble in a sea of black.[14] The biota has conspired and networked deeply with Earth's abiotic aspects, embedding its actions so seamlessly and deeply into the physical surroundings that it becomes conceptually impossible to think of life adapting passively to given surroundings. Life actively maintains the conditions suitable for its flourishing.[15]

Waves of Action

In 2013, the French philosopher of science Bruno Latour delivered a succinct interpretation of Gaia theory during his Gifford Lectures. In these lectures, Latour describes the relational interaction between different beings on Earth as "*waves of action* that do not respect any traditional borderlines and, more importantly, that are not happening at a fixed scale." To him, such waves of action "are the real actors which should be followed all the way, wherever they lead, without sticking to the internal boundary of an isolated agent considered as an individual inside an environment." They are the "real brush strokes" with which Lovelock has painted the face of Gaia.[16]

And aren't those waves of action a beautiful metaphor? Remember our fireside story in five acts: Whales will hunt squid and krill in the dark depths of the sea, then rise to fill their huge lungs with air. As they do, they defecate, up there in that marginal space where sunlight penetrates the ocean. Energized by the sun and fertilized by whale excrement, thousands of species of single-celled plant drifters will flourish and multiply. From there, waves of action ripple through into the entire food web: Zooplankton eat the phytoplankton, small fishes and crustaceans eat the zooplankton, larger predators eat smaller ones, and so forth. Algae send out distress calls to birds, some of whom spend years out at sea without ever setting foot on land, and the birds will hunt those who threaten the algae, and they will add their own excrement as fertilizers that will be more food for the algae. Meanwhile, clouds rise above the ocean, seeded by those same distressed algae under attack, and from way up in space, the tiny blue pearl will gradually, but significantly, turn whiter, reflecting more sunlight back into space, in turn regulating the surface temperature even further! To a planet whose sun has increased its energy output by a fourth since life first evolved, any evolved waves of action able to buffer and absorb the increased heat stress will help the planet at large to remain resilient, responsive. Meanwhile, millions of salmon across the northern hemisphere migrate from the food-rich Arctic waters to the continental shelves, carrying the riches of the ocean to the landmasses: More shape-shifting, more waves of action, spilling not only from the oceans to the atmosphere, but also from the oceans to the land, rippling through riparian communities as land animals, plants, and fungi feast on salmon flesh.

More waves of action roll round and round the Gaian turmoil: The permafrost grasslands of the tundra store huge amounts of carbon in the form of peat. This makes them immensely important for Gaia's temperature regulation. As it turns out, the tundra also has a high albedo, which simply means that it has a comparatively bright surface. So not only does the Arctic tundra reduce the greenhouse effect by storing huge amounts of atmospheric carbon in the eternally frozen peat, but the landscape also reflects extra sunlight back to space. It's a self-amplifying, positive feedback loop.

Muskoxen, reindeer, caribou, horses, and bison are the large grazing animals of the Arctic tundra. When their herds trample across the snowed-over plains, the trampled snow cannot insulate the ground as effectively as an undisturbed, thick layer of snow. This trampling could lead to a drop in soil temperatures, which would mean that the carbon will remain more firmly frozen in the ground.[17] In the absence of large herds of grazing animals, larger plants can also spread, including mosses, shrubs, and the firs from the great boreal forests in the south. As they do, the landscape becomes darker: The albedo drops, less sunlight is reflected back to space, and another feedback loop can kick in: More shrubs absorb more heat. More heat will hasten the thawing of the ground, which will allow more shrubs and small trees to spread their range northward. Peat will begin to decompose, releasing the stored carbon back into the atmosphere. Oswald J. Schmitz of Yale University and his colleagues have estimated that at the low end, the released carbon would be "equivalent to 10% of the fossil-fuel carbon emissions from China and the USA." At the high end, they suggest that "carbon release could rival annual emissions from China and the USA, and exceed Russia's emissions by a factor of 4."[18]

Each living creature and each ecological community triggers distinct waves of action that roll throughout the biosphere, ripples that interfere with other waves sent out by other agents, stirring local vortices and flowing into larger streams and currents, brushing individual lives, moving through food chains, and reverberating into the chemical composition of the atmosphere or the oceans, or the movements of tectonic plates. Each change or movement sends out its own wave patterns, moving other aspects of Earth's animate roundness. Each organism affects its neighbors profoundly. When many organisms engage in this process, each one trying to shift things for its own benefit, the collective result is a biosphere. In this understanding of Earth, suggests Latour, our planet can simply no longer be understood as something that has been "built" or "designed." There is no outside cause. The Gaian "anomaly" has brought itself into being, and it continues to bring itself into being actively from within. There is no outside source, no maker, no engineer, no designer creating the world from outside. Neither is there an intelligence within that intends to make the world in a certain way. With the exception of Earth's deep molten core and the black space beyond the thermosphere, everything is brought to the foreground. Everything acts. Everything "becomes a mediator adding its grain of salt to the narrative."[19]

Latour calls Lovelock's Gaia "probably the most secular entity ever produced by Western science."[20] If *secular* implies "involving no outside cause or spiritual basis" and therefore "fully of this world,"[21] then the metaphor of a self-composed planetary whole is indeed fully secular. The living Earth is the sum total of all these living, active forces that spill as waves around the globe, weaving the delicate web of life from near-infinite loops and cycles of participation, each of them a fine thread that adds another stratum of complexity to the self-birthing drapery. *Self-birthing* here is also a rather accurate synonym for the more technical *autopoiesis*, which is common in Gaian literature.[22]

Together, these waves of action compose the spherical membrane that is the biosphere, or the entire sphere in which life unfolds. The lifeworld of which phenomenologists speak becomes all of this: the rocks, the air, the water, the biota. Everything we encounter when we are wandering in the world is not just a passive set of objects. It is a dynamic interchange wherein the hydrosphere, the atmosphere, the lithosphere, and the numberless plants, animals, fungi, and microbes all interact with one another, spontaneously giving rise to a planetwide process of life. Throughout the complex, interweaving layers and scales that constitute the Gaian whole, life continually improvises itself into being. It organizes itself without script or design, never finished, never complete, never already existing. It writes its own story and creates the very conditions for its own flourishing. And every breathing creature, every chemical agent, every region of the seascape, the airscape, or the landscape, all contribute their agency to the living Earth's continual self-birthing. Nothing is just passive. Everything imbues its subtle influence on all else. And all are held within that strangely spherical body with its particular gravitational force. Their shapes, their bodies, their movements fit into a certain region of that sphere, and when it turns out that the planet's gravity is too weak to hold on to its water, the agencies conspire to do what a mechanical planet Earth alone could not: pull water back toward its surface regions.

Another way of saying this is that Earth is not so well understood as an exterior. Earth is imbued with a genuine interior dimension. Merely to study the exterior properties of the component parts does no justice to the emergent phenomenon that is this animate waterworld. With Gaia theory, the pre-Copernican and pre-Galilean experience of living inside a wombish world once more makes sense. But this time, the primordial embodied experience is fully resonant with the latest discoveries in contemporary science. Gaia, or Earth, is qualitatively different from anything else humans know about in the cosmos. It is no longer plausible to liken the living planet to a *res extensa*. Instead, we can think of Earth once more as a great inside, suffused with life, creativity, agency, sentience. Latour calls Lovelock our time's Galileo. The implication is clear: The scientific discovery of a living planet directly challenges the deepest assumptions and most inalienable truths of the Cartesian narrative tradition.

The River Within

The Anglo-American Nobel laureate T. S. Eliot penned the following poem in honor of some of the other eloquences who reside within that more-than-human world:

> *The river is within us, the sea is all about us;*
> *The sea is the land's edge also, the granite*
> *Into which it reaches, the beaches where it tosses*
> *Its hints of earlier and other creation:*
> *The starfish, the horseshoe crab, the whale's backbone;*
> *The pools where it offers our curiosity*
> *The more delicate algae and the sea anemone.*
> *It tosses up our losses, the torn seine,*
> *The shattered lobsterpot, the broken oar*
> *And the gear of foreign dead men. The sea has many voices,*
> *Many gods and many voices.* [23]

The sea, that *unknown universe*, as the German novelist Frank Schätzing calls it, is an inscrutable other, nearly unfathomable in its depth and expansiveness.[24] The sea is innately mysterious. It might cover more than two-thirds of the planet's surface, yet when I encounter it, standing by the ocean shore where foaming waves come rolling in, when I try to comprehend that vast Other that spans out there in front of me, the sea reveals itself mostly as an absence. As my gaze passes across the vast watery expanse, I become keenly aware that the sea withdraws: Try as I might, my gaze will bounce back from its blue or black or green or white surface, unable to be drawn in. To the creature of the soil that I am, the sea is predominantly flat, a two-dimensional disc. *Sea level.* Ground zero of my spatial orientation.

The sea shows times of great turmoil, when white-capped waves march through raging storm nights like a battalion of soldiers on a raid, or thunder charges through the electric sky like atmospheric earthquakes, but all that unrest merely stirs the surface. The expanses that lie beneath remain hidden, tentative, quiet. They withhold themselves from my awareness. I am not of them. It is this other depth, the depth of being above the sea and within the atmosphere, that is readily permeable to me. It is here that my ape senses most competently locate me. Here is the depth I most easily inhabit, the depth of the land, of being held in that peculiar gravitational embrace that steadies my feet and my gaze alike. Here I experience a resonance between the terrain and my senses, a poetic involvement, a somatic association between my own body and the depth of the living land. The sea, on the other hand, remains what it is: flat, inscrutable.

The sea has many voices. Many gods and many voices.

The Heidegger student Hans-Georg Gadamer has said that phenomenologically speaking, we participate in horizons that are neither entirely unique to ourselves, nor fully quarantined from others. Finding myself thrown into a horizon of reciprocal relations is to ask for the eloquence of others to make itself heard across membranes. What did Eliot say about the many voices of the sea? And what would it be like to read Eliot's poem phenomenologically? Might there be hints here for gathering our senses, such that even those who live inside this vastly other medium might begin to make themselves heard?

It tosses up our losses.

At times the sea shows itself as nothing but surface or exterior, casting the observer's attention back like a mirror, allowing nothing to sink in. Remaining silent. Then again, the ocean might make bold statements, speaking unexpectedly and suddenly, with exclamation marks, *tossing up our losses.* A torn seine that bops to the surface, a shattered lobsterpot. Gear of foreign men. The bodies of foreign men. There is a strange double motion at play here. The perceptual horizon of the ocean surface prompts its own strangely two-directional pull, withdrawing and simultaneously oozing forth allusions, clues, intimations, all the while engaging the inquisitive ape who's standing there on the shoreline in a dialectic of ebb and flow. The ocean withdraws, then tosses something up; it withholds, then exposes. The ocean presences itself as an enigma, with a clear and distinctly other-than-human sort of agency. It will not be dictated. Its surface acts as a boundary that both connects and separates, uttering a semantics of transformation: The seine is broken, the pot shattered. Foreign men have perished and been committed to its subsurface silence.

It offers our curiosity.

In offering (rather than demanding), the ocean holds us in a participatory stance. It is a participation that is far from seamless. The privacy of the subsurface dwellers remains as resilient as the privacy of those who live above. Each is deprived of their unbroken participation with the other. None will be fully exposed; each has only the horizon of the ocean surface to participate with those on the other side.

As a philosophical term, the notion of the *horizon* has been established since Nietzsche and Husserl first popularized it, before Gadamer later made use of it. As a word, it derives from the Greek *órízon* (ὁρίζων) and means "the bounding circle." Horizons are the permeable boundaries at the edge of perception. Gadamer thinks of them as passageways, porous membranes that connect the depth of our own, localized center with other depths of other localized centers. Horizons are the skins that allow for uniqueness-in-relation. To find ourselves thrown into a horizon of reciprocal relations is to allow for the possibility of reciprocity across boundaries.

Thinking is finite, for it is necessarily located within one's own horizon, but because horizons are not hermetically sealed, there is always the possibility of setting oneself in motion and casting a light onto what lies beyond. As there is no end to this motion, finite, embodied thinking has access to infinite possibilities to learn.

Gadamer uses the much-quoted term "fusion of horizons" (*horizontverschmelzung*) for this qualitative exchange that happens as two horizons encounter one another in their porousness, each illuminating with their radiance those regions within that other horizon that are briefly brushed by the encounter. Importantly, this fusion of horizons never results in a single, unified horizon; it never compromises the particularity of the distinct horizons that encounter one another.

The work of fusing horizons is a task that cannot find its own completion. It must remain tentative and fumbling, suspended in a dynamic tension with those other horizons that we slip across as we move through the world, illuminating some of its regions, casting shadows on others. Each fusion of horizons is a relocation. Each fusion is a reorientation in the world. Each illumination darkens some aspects of the world. We live inside horizons that steadily move along with us.

Being human, fully of and within this world, we find ourselves thrown into a horizon of reciprocal relations. And because we have no access to a view from nowhere, there can be no absolute knowledge. Our minds are continuous with our bodies. We radiate across the membranes of our horizons and outward into the terrain, where we will quickly find that there are other sentiences, each of whom guards their own set of horizons. Knowing, for the human animal, becomes a function of relating. It is at this fluid boundary of our perception that the ocean makes its offerings to our curiosity and nurtures our sense of possibility.

It tosses its hints of earlier and other creation.

Consider the following possibility: The ocean surface is not one horizon but two. The first is the more familiar. We speak of horizontal extension, by which we mean the way the Earth curves away in all directions at the outermost periphery of our awareness. If phenomenology enacts a way of being in the world that is participatory, then to approach the world phenomenologically means to set oneself in motion. Moving toward the horizon will reveal certain aspects of the terrain and conceal others, as some rise before me from below the horizon ahead and others sink back into the horizon behind me. Being a spherical body, the planet bounds, or limits, our ability to know it.

Then there is the other horizon. Unlike the first, this horizon has a vertical extension. On the ocean surface, two very different mediums meet—the atmosphere above and the ocean below. As a human animal, my body participates most freely and with the greatest subtlety in the depth of the atmosphere; the world below this vertical horizon, below this boundary, remains far less accessible to me. And yet, I wonder: What would it be like to encounter those who dwell on the far side of my perceptual horizons, there inside the oceans?

We cannot hope to make these others fully available to our scrutiny, for they will withdraw. Such is the nature of horizons. Ignorance is not an epistemological problem as much as it is an aspect of being enfolded in the ever-changing fabric of the living world. *The sea tosses its hints.* Hints are all we can relate to.

And yet we strive to speak and to interpret the hints that are tossed our way. Recall that to Heidegger, it is first and foremost through the poetic that nature is disclosed to us. Heidegger thinks of *nature* in the original Greek sense of *physis*, or "that which emerges from within itself" (*das von-sich aufgehende*). The earliest documented occurrence of *physis* is in the *Odyssey*, where Homer uses the word to describe a plant's intrinsic growth pattern as that plant unfolds into the open from within itself. Nature, in this primordial Greek sense, is a process of *autopoiesis* or "self-poetry."[25] It is a dynamic and ongoing emergence of Being, experienced as having that self-birthing quality we are now rediscovering through the scientific image of Gaia. That is why this primordial experience of the self-birthing of Being was later translated by the Latin term *natura* ("birth," from Latin *nasci*, "to be born").

Heidegger seeks to rejuvenate this original, animistic understanding of nature. He rejects Western philosophy's metaphysical interpretation of Being as a solid ground, as eternal substances; his thinking discloses Being as spontaneous, localized, particularized self-emergence.[26] As we saw earlier, Heidegger thinks that it is through the poetic that "everything first steps into the open."[27] Poetic speech is that craft which "brings forth" the world and bears witness to the world's self-birthing. Poets such as T. S. Eliot hold a place of special significance in rejuvenating the experience that the world speaks, and in refreshing that experience with original metaphors for those who have lost it. Poetic speech is a crafting of words, the continuous effort to place the ephemeral and fluid landscape of the imagination squarely within the expressiveness of multiple agencies inside the articulate Earth. Poetic speech stands guard at the boundary of two linguistic communities, at that fluid horizon where the chatter of humans meets the many-voiced, subtle speech of the more-than-human community. From this precarious place in between, poetic speech slips in and out of both worlds, listening, observing, bearing witness, receiving words, offering words, searching for meaningful exchanges between these worlds. Poetic speech becomes a bearer of intelligibility across boundaries. It is a work of translation and of transformation. It is a work of keeping alive the animistic experience of a world that speaks. To those of us who live more squarely within a purely human discourse, such poetics offers to hold the awareness open to that larger chorus of voices out of which our conversations rise, and into which they need to return for nourishment.

The river is within us, the sea is all about us.

Eliot's poem is a lucid recognition of two potent and closely related philosophical phenomena. The first is that of horizons. The phenomenon of horizons lets us treat epistemological boundaries as malleable and semipermeable membranes, rather than hermetically sealed borders. And although Gadamer himself reserved the term for strictly interhuman discourses, it is not difficult to see how the term is relevant for us:

To think in terms of epistemological horizons allows for the possibility of other eloquences to inform the thinking mind. Salmon articulate themselves in ways that are unique to them, using expressive means one cannot easily translate into hypotheses or propositions. And yet given the wealth of our understanding of salmon, we must assume that they, too, possess a unique consciousness, and that they, too, inhabit this Earth from within a horizon that is particular to them.

Is it possible to bear some of the salmon's expressiveness across the horizons that hold us apart? Is there a way to convey their fluid, sinuous relevance in ways of speaking that resonate within our own embodied thinking? It appears clear that we must approach the epistemological horizons that hold us apart with the utmost care, with humility, choosing our steps judiciously. This requires a certain patience that must be coupled with a readiness to listen at length before we venture to speak on their behalf. It would be naïve to insinuate that one could simply dissolve the horizons that are between us. It seems more realistic that any meaningful participation would succeed in spite of horizons, in attempting some kind of fusion. Each of us is an embodied center. Each comes to know the world through the subtle and omnidimensional flow of reciprocal participation endemic to the Earth region wherein we live. Salmon decidedly partake in what Keats called the "poetry of Earth"; they speak, but in ways elusive, and weird, and not at all easy to transform into human styles of cogitation. A certain inscrutability seems immanent. When we venture toward that horizon, we must do so, as Abram has suggested, with a language parsimonious and precise. We must strike a balance between what we rationally know about their lives and what our bodies allow us to intuit, feel, and sense.

The second phenomenon resonant with Eliot's poem is the world's depth dimension. "Depth" refers to the way in which we are given to the world in a necessarily ambiguous manner, with some things hidden and others close at hand. The monocentric story of humanity-as-separation ignores the primordial way in which humans, too, participate in the world's depth. Abram writes of depth as that which "implicates the whole of our animal body (this carnal density of muscles and skin and breath), situating us physically within the animate landscape."[28] To turn the attention toward this depth dimension is to rejuvenate an experience of the world as a great interiority, as something that we are inside of.

Because I wholly participate in a much larger and more diverse inside, I cannot see everything at once. Some things are hidden from me, while others reveal themselves relative to where I stand in the depth of phenomena. Contrary to the basic modern assumption, there is no *res cogitans* that can get a complete view of the world. I cannot see the world from every direction at once and assume that the world can be completely and objectively understood. If mind really is inseparable from the body, and if the body is situated firmly in the depth of the ecology, then it becomes quite impossible to speak of the world in purely objective terms. This is a world I experience only from *down here*, from *inside*. My own perspective remains *a* center of the world, but the monocentric epistemology is shattered and broken, giving way to an ambiguous and never fully conceivable multiplicity of centers. As Abram describes it: "We are many sets of eyes staring out at each other from the same living body."[29] A more accurate description of the world might emerge from asking, how precisely do I stand in relation to these other beings?

Because I am a carnal intelligence, with a mindful body fully immanent in the depth of the more-than-human Earth body, alongside sperm whale, rainfall, tidal patterns, leatherback turtle, and migrating salmon—because I am situated firmly inside this atmosphere, looking up at these clouds, balancing on this rocky mantle of Earth's crust, I am able to recognize the speech of things even when these things articulate themselves in ways vastly unlike my own verbalized thinking. Abram describes how the recognition of our immersion in the land's depth entails a certain fluid understanding of the self and of the world. The former subject-object division is dissolved, as all that was conscribed to the background is now rigorously swarming into the foreground.

The theologian Douglas E. Christie, reflecting on Abram's work, takes an interest in the possibility of exchange between our own, unique bodily selves and other sentient bodies, those potent, rich encounters that allow us to recognize the fluid and porous quality of the borders between ourselves and others, between humans and other-than-humans. He writes of the importance of working to "relinquish our imaginative attachment to boundaries and hierarchies that keep things distinct and separate and to reimagine a world that is fluid, relational, organic."[30] "It seems clear," he goes on to say, "that we must discover again the kind of expansive, fluid categories of thought and practice that can help us understand what it is to feel the touch of the other, to enter into the life of the other."[31]

So what would that actually be like, for Earth to experience itself not from a human angle but from another? What would it be like to surrender the Cartesian dream of reflecting upon a material world as if we were pure minds outside of it and instead recognize our carnal intelligence as an attribute fully immanent in and participatory with the biosphere at large? What would it be like to allow our perceptual horizon to become so porous and permeable that other embodied agencies begin to speak from beyond the porous seam of our own bodies? To actually strive to fuse horizons?

CHAPTER EIGHT

Being Salmon

*Ardor: the Earth's fervent song, which we answer
with our own, imperfect song.*

ADAM ZAGAJEWSKI

*A*ttention is the currency in which relationships are forged and rejuvenated. She pays attention to the others. Her eyes, her tongue, her lateral line, her muscled fin, her magnetite-rich skin—the confluence of her senses flowing back and forth between her nervous system, the others, and the encompassing river.

Her journey began several moons before she made it downriver to the brackish waters of the estuary. There was a night she knew the time had come. The year's spring floods had passed, and the water was beginning to warm. It was the night of the new moon, and the sky dome was black, the first seamless darkness after a stretch of moonlit and starlit nights. Three summers and winters had passed since she first emerged from the gravel. Three summers and winters, and never in those gyrating seasonal turns had she strayed very far. She and all the others had staked out their tiny river bottom territory, and they had defended their small home ranges jealously. But now the rising temperatures and the heavy gray sky beckoned. They struggled to make contact with her. It is time, they said, each in their peculiar tongue. And though she had not heard these utterances spoken before, her body understood. She abandoned her plot, watched the current wash away earlier rivalries, and joined the tight school that was beginning to crowd together. More and more of them arrived, and her body diffused, mingled, dispersed among the others, until they all had morphed into a larger, mutable shape. They were a single, collective will with a multiplicity of watchful eyes, sensing an urgency pulse within, a certain rush.

A cloud shadow passed overhead. Perfect darkness spilled through her gills, her breathing canals, her eyes. This was it. The many-finned, pliable body seeped into the fast-moving torrent, tails downcurrent, heads facing upriver to breathe the water while the river carried them. She did not struggle to keep abreast of the current, or to move against it. The larger body she had become diffused further outward and into the larger body of the water itself, becoming its current, its resolve. She abandoned herself entirely to its guidance, breathing it, letting its drift become the measure of her imagination.

Thick, black, watchful water. She could not see the riverside shadows passing by above, but she knew they were there. Towering columns in the expanse of sky, all bending toward her narrow stream from either bank, their gnarled limbs reaching up, coaxing eerie voices from the high air, whooshes, wails, surges, howls, rustles, crackles, moans, shrieks, gnarls. Sky-rivers, wooden and ancient: Each the precise timbered image of a vertical watershed, nourished by air, ever pouring themselves out of the atmosphere and down into the soil. And among these concealed sky-rivers, smaller, quicker shadows: featureless faces turned streamward, ears jerking, beaks poised to strike, noses twitching, tails erect, wings ready to take flight and scoop. Lungbreathers. All examining the passing water intently. All eyeing, earing, sniffing toward her.

The days came and went, and she drifted, mostly under the cover of darkness. Smaller passageways joined the central river vein, water that tasted almost familiar. Scent ribbons bled through the turbulences, not unlike northern lights that bleed into a winter night. She smelled and smelled, and somehow, it seemed that the world was deepening, growing larger, there in the margins of her awareness. The scent ribbons wound upriver, all the way back to where she first had left.

Changes were adrift deep inside the veins of the watershed, and deep inside the fabric of her flesh. She intuited the changes. Her body, growing longer and slimmer. Her steady impulse to flow, to move and be moved. Her fin edges darkening, turning shadow-black, the fins themselves growing more and more translucent. Her very skin changing, turning silver, looking ever less like the turbid river and ever more like . . . well, she was not exactly sure like what.

She had noticed the signs much earlier, even before this journey began: the vain attempts by some to form schools when most were not yet interested. The occasional flash of premature silver skin among them, when all the rest where still river-shaded. The larger and more slender body that so clearly stood out among the rest—the last thing any of them wanted, now that this school of duplicate bodies was their only refuge. Each of them had been on their own trajectory. Each of them had intuited an imminent metamorphosis, but no one was quite able to make the loose ends flow together. They knew changes were adrift, but they were slightly out of sync with one another, out of phase. Isolated in their own skin. Until the new moon and the warmer water incited in them a common pace and purpose. They knew better than to resist submitting to these eloquent powers. And so, she and the others drifted, and they felt for the clues. They labored to become fluent in the subtle tongue of synchronicity.

Others like her spilled from the smaller tributaries. Some drifted downstream with their tail first like herself, their noses pointed home-ward. Others dashed past her headfirst, in haste. Still others seemed to swim lazily upstream but were outpaced by the water's velocity. She blended in and out with the others like the different waters here, and schools shaped and split and reformed. There were sudden outbreaks of anxiety when the swarm swiftly imploded into a dense, flashing orb. There were times when they relaxed and loosened into a scattered cloud, a *seeing* cloud. Marauders tried to break their lines. But often they left again hungry, discouraged by the tight and yet intangible ball of a hundred silvery flashes, jolting and jagging erratically before their eyes, their resolve to eat frittered away in the frenzy.

This was not her first arching journey. Way back, in the beginning, it had taken her less than a day to complete her first minimigration. She writhed her tiny body from between the pebbles, and she kicked and kicked, and she wriggled from the river bottom and up to the water's edge, stayed there just as long as necessary, and then hurried back again. It was treacherous but inevitable. She had consumed nearly all that was left in her yolk sac. It was either take the risk or starve to death. It was the first time she felt that ravenous appetite, that single-minded focus on food. She was hungry, and so she had to go to the boundary of her known world. When she got there, she opened her tiny mouth, and she inhaled a deep gulp of air. The air gushed into her, and she felt its small turbulences gather within and congregate down along her upper back. She became *buoyant*. A breath of air, and she was weightless, able to fly in water. She was ready to hunt! Her first true migration, her first metamorphosis.

How long had she been drifting? Two full lunar cycles? More? One night the current began to slow at last. A new awareness began to pour across her tongue and through her gills. She tasted it, breathed it. Part curiosity, part faint memory, part anxiety, it dawned within her flesh like the great luminary of day, as when the gleaming sphere rises from the water in the morning, pushing the curved night-shadow overhead toward the far horizon, breaching the darkness like water breaches a massive logjam, pouring out light. Her shuddering muscles, her fins, her scales, the whizzing network of her nerves—her entire pulsating body was strangely on edge: *salt!*

Here it was, the answer to a question that had been rising steadily from the abyss of her sentience: All along, she had been in preparation. Each novel sensation, each morphing passion, shape, and appetite, all were an orchestrated flow of intentions, passing back and forth between the river and her body. The closer she came to the estuary, the clearer the determination: to leave, and to survive the extreme change as she was now about to journey into this vastly different and larger body of water, the sea.

Metamorphoses

Are we overanimating the salmon? Are we creating some comic, anthropomorphic projection of what it is like to be them? Or are we beginning to allow them to show themselves in their agency? If parts of the narrative appear fantastic, then surely not because of the unruly wildness of my imagination, but because the experience of being salmon is fantastic in itself—even if it tells itself tentatively and incompletely, even if it never exposes itself to us entirely but holds aspects in concealment, even as it slips imperfectly across the semipermeable horizon of our human senses and is shaped into forms resonant with the localized and limited eloquence of human speech.

What would it be like to be a salmon, to cast oneself resourcefully and alertly into the whirling web of the Gaian biosphere from *over there*? Such a question comes knocking as soon as we step outside the Cartesian dualism and acknowledge that others act as sentient minds or thinking bodies. Recognizing that each of us is situated inside a semipermeable horizon that is not confined unto itself, the question becomes: How can we meaningfully encounter one another at the outer edge of our perception? Gadamer said that "a person who has no horizon is a man who does not see far enough and hence overvalues what is nearest to him."[1] Such a person becomes self-centered, believing themselves at the very center of the world, believing that their style of sentience exhausts all the possibilities for sentience inside Earth. Horizons help us break with that monocentric superstition. They multiply the number of centers from which the world experiences itself.

Biology, ecology, and behavioral science have come a long way in recent decades. They have at last begun to càtch up with immediate, bodily experience. Many with direct, experiential knowledge of other animals were never convinced that their animal companions did not have a world. They were never convinced that animals should be stuck, as it were, in a determinate and closed circle of an environment. There is no longer any empirical ground for believing that only humans can have a world, and that all others dwell inside a closed circle of an environment. Environmental freedom is a definitive characteristic of any living being, no matter whether we are thinking about humans, salmon, fungi, trees, or microbes.[2]

In Latour's reading of Lovelock, nothing is entirely inert or mute. Everything acts! The world that other-than-human living creatures encounter is not prematurely unified or closed. It must rather be composed through an infinitely complex participation with other agencies. Every agent exerts an inimitable poetic quality on every other agent, causing resonance, repulsion, consonance, dissonance; causing a reaction; sending a wave that spreads out and joins with other waves; setting the entirety of the animate globe in motion as it flows round and round in fluctuating loops. To find oneself alive amid such a vigorous play of powers is to continuously fine-tune one's movements, choices, or actions. It is to adapt to and conspire in creating the conditions necessary for flourishing.

Abram's treatment of the phenomenon of depth helps localize the embodied mind concretely inside places and inside the patterned flux of things. Earth in its depth becomes a composition of overlapping provinces and localities, each of which originates distinctive modes of thought and action, different styles of being-in-the-world, different sentient bodies that coevolve in creative participation with the other agencies composing that region. The depth dimension implies that if we wish to even begin to think what that must be like, being salmon, our thought must resonate both with our contemporary understanding of the salmon's physical transformations and with the effects of Earth's particular atmosphere, its gravitational pull, its magnetic field. These should not be treated as "facts" to dress up the story. They are not decorative props in the background, used to fabricate a certain "narrative environment." We must ask to what degree these factors are relevant *in their agency*. What are the agencies that compose the oceans? Abram suggests that this intercorporeal life of the world cannot be determined by a set of objective (i.e., "factual") processes; this life is fundamentally *metamorphic*![3]

Only by attending with heightened alertness to our own bodily participation in the world's metamorphic depth can we begin to ponder what it is like to be salmon. By inhabiting our bodies more attentively, we will be able to listen, with Abram, "to the sensuous play of the world, allowing the unfolding pattern of that display to carry us into a place of dark wonder and possibility."[4] This is how writing can both be awakened by our senses and awaken them in turn. Our journey into the sensuous horizon of the salmon is not a literal one, nor is it mere metaphoric speculation. We are exploring the possibilities of what Abram thinks of as metamorphic speech:[5] speech that thinks not "about" an "object of study," but that strives to participate in the continuous shapeshifting of the palpable, sensuous world. Metamorphic speech: to encounter, from our own place inside the depth of the biosphere, the salmon in their own depth, over there.

Thinking Like the Ocean

To be thinking like the ocean. To be the ocean thinking itself within her. Perhaps this is precisely what it means to reach maturity. She is called into being within the fluid depth of water, which is at once elder, womb, blood. How long has her kin voyaged the arching globe? Six million years. The steady stream of ancestral journeys reaches back into time immemorial, and their shapes are slowly morphing, changing, reworking themselves across this vast curvature of time. She lives inside an imagination that has been molded by glacial advances, by the patient force of trickling water carving deep flanks into flatlands, by a trillion raindrops eroding the mountains. Her sentient body bears within itself the promise for metamorphosis, a creative adaptability within a world that never rests. As she grows, the distant echo of a particular watershed takes shape within—its velocity, its seasonal temperaments, the power of its autumn swells, the complex topography of its arteries. Each quality of the river adds its subtle claim to this body that is her, refining her. Born into a shapeshifting world, it is what she is: a shapeshifter. Swelling rivers, marching glaciers, dwindling mountains, currents that flow on and on, the very ocean: Each remembers itself within her flesh. Each calls itself into being through her flesh, again and again. She is the world birthing itself.

　　She knows nothing of the furnace deep within the core of the world, of pressure so immense that molten iron will crystalize into a solid. Nor of the liquid iron that flows in a rotational pattern around that innermost core, following the planet's rotation. And yet she can sense the delicate magnetic bands that weave themselves from these frictions and outward, around the spinning axis of the globe, fluctuating most forcefully near either pole and weaving smaller, far subtler bands between there and herself. Earth's bipolar magnetic lure flickers continually within her. This globe's composition and its massive shape rebound throughout her flesh. That far larger body throbs in her head, all along her lateral line, throughout the varied topography of her flesh. To align herself with true north is to sense a faint, subtle shudder of recognition rush through her. A chill of embodiment. Iron crystals within her, iron crystals within the core of Earth's larger body: beckoning, striving to hear, calling, responding, gesturing, learning to react, aligning themselves, seeking congruence. Ever since she left the river, she has been negotiating the fluctuating semiotics of field navigation. As she has matured from a smolt into an adult, her sense for the larger body has grown keener. Each local variation in the blue expanse—its own field quality. Each region in the ocean—its own magnetic tension. With sustained attention, and if she engages the full range of her corporeal intelligence, she can distinguish the unique feel of the magnetic field where she first encountered the ocean as a juvenile. No other place sets her nerves on edge quite like that one.

Her hatchery-raised brothers get frequently lost at sea after they escape. Raised in a world of electric wires, electric magnets, and swarms of metallic objects, and driven mad by circling endlessly in cramped pens, their navigation skills become crippled. It does not help that so many of them bear the marks of captivity. There are those who suffer chronic cataracts and fall blind. Others have blunted noses. Others suffer from humpbacks, incurred from high temperatures during incubation. Others have fissured fins and tails from fighting and from rubbing constantly against the cage mesh. Their tails and fins are also smaller, for besides circling the pens, the only exercise they have is in leaping, typically in vain, to rid themselves of lice. Even when they do escape, they are not as likely to learn to recognize the distinct signature of home. They have not learned to perceive themselves as enfolded by the larger estuary body, and by the even larger Earth body. In some sense, they are deprived of the possibility to fully mature.

She and her kind are not the only ones whose sensing bodies can extend from here to the magnetic poles and into the very core of Earth's body. Others can do it: stingrays, sharks, lobsters, loggerhead turtles, certain aquatic microbes. Above the surface: many of the winged migrants. On shore: foxes, deer, cattle, flies, bats, honeybees, and yes—we two-leggeds. All carrying within ourselves traces of magnetite. All able, in principle, to probe attentively into the kind of embodied chill that she has felt. All, like her, able to learn to locate ourselves within. In vastly different ways from her, most certainly. But within, just the same.

The ocean is never at rest. The waterworld she inhabits is constantly on the move. She rides huge currents that carry her far offshore. Many of those she has come to hunt also undertake vast migrations into the blue unknown. No script can predict where they will be at any given moment, and how she will find them, and what she will do once she does find them. She cruises the currents, she observes the cues, and she makes choices. She senses at all times the direction and rate of moving water. She senses her own movement relative to the water, and that of predators as well. The turbulent vortex of a fleeing capelin is as audible to her hearing skin as clouds of unsuspecting shrimp, or even kelp floats and other debris that drift sluggishly at the surface. Each leaves a trace imbalance in the pressure that flows around her, rippling waves that brush along her lateral line. For her, to be touched by water must be a kind of hearing, with the whole of her body. Hearing, with any sensuous receptor organ, is a way of interpreting pressure changes. For her it must be a kind of full-bodied eavesdropping, as distant, faint thunder rolls through the innermost regions of her body. She is stimulated by the pulsation. She resonates with the touch of sound. Minute hairs in her skin are bent by changes in water pressure, and they trigger a recognition. Her hearing skin extends her sensuality into the blue expanse. She thinks herself into the soundscape.

I try feeling into the thickness of the sensation: strata upon strata of vibrations, waves, tremors, pulses, overlaying one another, swelling, spilling, surging around her and through her. The blue whale's thousand-mile chant; the shrimp's industrious click; a huge cloud of herring, panicked and driven to insanity as dolphins make deadly and impenetrable walls of bubbles whiz and rise around them; solitary sperm whale, ascending like an apparition from the deep; bluefin tuna, dashing alongside four hundred fast-moving dolphins; a fierce Arctic storm that agitates the abyss, making clouds of nutrients rise to the surface, promising to her a moon of abundant hunting; feathered ones who come diving down in a boil of foam and bubbles, beaks first, wings folded to minimize impact, spreading their wings and whizzing through the water, chasing her. Each brushes her skin, each calibrates her spatial orientation, each lures her sentience from the porous confines of her body and into the ocean around. Her awareness gushes ever outward to suffuse the four-dimensional depth. Her mind is in her and it is in the ocean, so much so that the ocean does indeed think itself in her.[6] Each successive sound and pressure change bellows and bowls through the water and jolts through her body, inciting her. Each spark sets some region of her awareness ablaze. At times the sparks come flashing in such rapid torrents, from every angle, from near and far, that every fiber within her feels utterly irradiated. Her sentience attunes itself instantly, effortlessly. She is awake, and she is alert. She knows exactly what to do.

When navigating the ocean, timing is crucial. There are no landmarks to steer by, no fixed places to commit to memory, no permanent certainties to help her locate herself, and so she turns toward all that moves in patterns. She takes cues from fluctuations in water temperature, from wind patterns, from the journeys of all bodies—marine, terrestrial, celestial. Among these, none gives her guidance as clearly and radiantly as the great fireball that hatches from the ocean every morning. She has long gleaned its cues as it migrates up into the sky early in the day, reaches its zenith, and then begins to return, only to leap back into the ocean again at dusk, moving in a big sweeping curve. The rising, juvenile sun is carried from one medium and into another, and as it journeys farther away from the ocean and up into the sky, it grows in strength until its body is sizzling and blazing with energy. Then it begins its homeward migration, homing in on the ocean below. As it returns, its colors change and spill themselves through the sky and into the water, spawning light. And then the sun perishes, and it rots, and night spreads, and a transformation takes place within. For where there is decay there is also rejuvenation. And sure enough, a new fireball will hatch the morning after, as determined as the last to set out and complete its own journey.

She has been observing the sun's passage for so long that its body—like the river body, like the Earth body—has slipped into her own. Even from below a turbulent sea she can gauge the precise angle of light beams cast by the sun above. The angle tells her exactly where the sun is. And even when clouds cover the sky, she sees the sun. Some light still diffuses into the water, and that light is polarized in a distinctive pattern relative to the sun's direction, even deep down the water column. Her body has learned to read cues a human observer would be blind to. But simply seeing the sun would not help her navigate. Locating the sun spatially would be meaningless for her, because all is in flux here, the sun included. She also needs to sense the sun's pace, to adjust for its movement in the sky at any given time. Only then can she use the arch of the sun's passage and calibrate her own position relative to the topography of the larger ocean body.

It might be tempting to think of an "internal clock," but the mechanical metaphor approaches the question from the wrong angle. It externalizes what is deeply participatory; it presumes not her style of thought but our own, which is more centralized in our head, and which has a way of broadening our perceptual horizon with the help of tools our hands have crafted. Her intelligence spreads with less discrimination throughout each sentient region of her flesh and extends out into the deep blue. Perhaps she and her kind have tracked the sun for so long, with so many watchful bodies, throughout so many life cycles, that the exact measure of its journey has begun to beat within their flesh. Perhaps for her, to navigate the ocean is to strive and tune in to the many layered pulses of Earth's continuous self-birthing, including the celestial bodies she sees arching across the sky. Perhaps she thinks in patterns of ebb and flow precisely because it is the lingua franca of her earthly, marine, and sky elders. Her intelligence is as ancient as her journey, and inseparable from it. But it is also, and in equal measure, utterly fresh: It calibrates itself constantly and fluidly as she and her kind slip from one life cycle to the next. Navigation becomes a creative dance between perception, the memory of her breathing flesh, and the larger bodies that compose her. It becomes a continuous improvisation with the upsurge of the present moment, in a complex field of interaction. What a feat! What a bizarre and admirable accomplishment!

Latour's work on dissolving the outdated metaphor of an "environment," Abram's work on the biosphere's depth-dimension, and Gadamer's work on horizons help us place thought explicitly into the tangible and unique atmosphere of this planet, these oceans, and this soil. Mind no longer appears to be an exclusive or private substance, but a kaleidoscope of ways in which the self-composed Earth has learned, through eons of evolution, to organize itself. Mind is an Earth quality. Mind is Earth that, over the course of 4,500 million years, has come to reflect upon itself through a nearly unimaginable variety of sensuous bodies.

If we were to give a name to the method that brings Latour's, Gadamer's, and Abram's writings into conversation, we might call it *embodied ambience astonishment.* What are the composed dynamics of the atmospheric envelopes that hold us each within? How do we and others participate in waves of action that originate elsewhere inside the Earth's living depth? How do we locate ourselves within those Earth regions that are endemic to our kind? And from that concrete place within, how can we more keenly recognize those other sentient beings who are within another region? If all things are in motion and each thing influences all else from within the shared, carnal field of experience in which we all participate, then our questioning must try to articulate as many loops of participation as possible.[7] It requires us to look for details both in what we know about being carnally embedded in Earth and in how we speak of what we know. It requires us to think, intuit, feel, and sense salmon not as discrete ontological units but in their relational aspects: salmon as sentient creatures, as patterns and perturbations, as creative agents internally related to other agencies. Salmon, like us, are fully of this living Earth.

Homebound

Accepting the responsibility of return is never a trivial matter. She faces a journey that she is likely to make only once in her life. Death is the rule, survival the exception. It is a journey against all odds, and yet she leaves behind the rich ocean-foraging grounds and turns back. She gives up on sand lances, capelin, crustaceans, squid, and all the other bodies whose flesh has become her flesh—an abundance so great that she has grown into a formidable hunter, 150 times larger than when she first swam from her estuary and into the great wide open. She is about to face starvation that might last a long succession of moons, at the end of which she will likely die from exhaustion.

She is bodies within bodies within bodies. She is a transient traveler, the world's momentary midpoint, the world pulled together in a throbbing core of heightened awareness. She is an inside within a larger inside within a larger inside. She is semipermeable to the others, receptive, reaching out through the various strata of porous membranes that connect them and hold them apart: the skin, the water surface, the vault of the night sky above. She is a unique event with a definite beginning and end, and she is the foaming crest of a wave that rolls steadily onward through the generations. Body and movement, a drop in the great river and the river's flow. It is simply who she is.

Journeys within journeys within journeys. She is one of the few who will complete the largest circle. Most are picked off along the way. Once, fifteen thousand miniscule translucent red globes lay hidden in the river bottom gravel, scattered throughout the many nests her mother had dug, each red globe rising within to its own tiny dawn. Yet many of them never woke from that dreamtime. They were discovered and devoured as their own alertness was beginning to stir, and they slipped back into the soil's mysterious alertness. Only a fraction of them set off down the river in the first place. With every tail-kick along the journey, her life was likely to be cut short. Every successful hunt also meant that she had cut the life of another fish or squid or crustacean short, taken them out of their personal life cycle. Every one of them had also strived to draw that perfectly round cycle around their story, but they participate in a vigorous struggle between striving and achieving. The content, the flesh of who "they" are fluctuates, at times in turbulences so stormy that she hardly attracts attention among the multitudes. And yet here she is, improvising her vast circle through the ocean, wanting to live.

And here she is, returning. The sun and the moon, the magnetic fields, they help her approach the coast at the right angle and at the right time. Once within reach of the great landmass, her phenomenal sense of smell is greeted by the hundred smells of fjords, streams, rivers, creeks. And so she searches for that one particular smell among the others, and when necessary she breaks out of a larger school of fish and joins a smaller group. She zigzags up and down the water column to find the water layer that smells like home. She does what is necessary, until she knows she has found her way.

When she enters her birth river, she stops eating altogether. For moons, she will swim against the current without ingesting a bite, and her life will be on edge. And yet she presses on, following the same scent ribbons that flowed here, way back then. When she loses the trail, she backs up until she finds it again. There is no time to lose. Urgency is visible in how her body prepares for what lies ahead, for what she cannot yet fully fathom. The males will develop a hooked lower jaw, a large and swollen upper jaw, and a flashy rusty-red skin. She, on the other hand, assumes a murkier, less conspicuous skin tone. Whatever surplus energy she has, her body directs toward a single, absolute cause: her roe. As she migrates up the river, she will transform up to a fourth of her own body weight into roe. It is a new kind of heaviness, a swelling urge to push out that which is growing there inside of her. The desire becomes more demanding with every day she moves upstream.

One day, she is back again, and others have arrived, too. And it begins. The males stake out claims in the gravel and meet any trespasser with sudden and uncompromising hostility. Violent eruptions are so common that the water will boil as if atop a volcanic sea vent. Meanwhile, she has nests to dig with her tail fin, and she must dig many. She never squeezes more than a few hundred eggs into any one nest, often as few as fifty. If calamity strikes, only a fraction of the new lives will be consumed at once. And so she digs, and she covers up the nests she has finished, and she digs again. She continues for three-quarters of a moon, even when her skin and flesh start to peel off, even when her sores begin to fester, even when her energies are free-falling, even when more and more companions beside her are discovered and torn apart by riverside predators. Each time she decides that the hollow is just the right depth, she will lower her tail end into it. A number of large males have been waiting for that to happen. She will make her pick among them, and she will allow the chosen one to lower his tail into the hollow, side by side with hers. Mouths open, the two of them touch one another and all suffering fades—the journey, the fatigue, the heaviness, the starvation that has intensified until her flesh has begun to consume itself. In this moment of ecstasy, she lets go of it all. And the eggs spill from her body and into the womb of the water.

He, too, lets go. His milky cloud veils the scene. Smaller males—jacks—will sometimes come dashing from their hiding spot, zipping past the two, and without slowing they will add their own contribution to the milky cloud. She simply acknowledges the jacks. It is as it is. This moment is what they all have come back for.

———

David Abram once pointed out to me that whether we are salmon or two-leggeds, whether we are feather-clad, barked, tentacled, or hooved, each of us fellow creatures has our own cognitive style, our sensory skills, our uniqueness. And yet, he said, each of us is also listening to the discourse of other things.[8] It might not be easy to interpret the discourse of other-than-human creatures, but it is not entirely impossible either. We partake of that common depth. We are of a common flesh and ancestry. Each of us is our own center, but we can relate to one another across the semipermeable horizons of our perception.

It might be that her body is not the only one that bears the promise for metamorphosis. It might be that metamorphosis is a faculty available to us all. To listen to the speech of others is to investigate very cautiously the shapeshifting qualities of our own mindful bodies. Not to anthropomorphize these others, to uncritically project human qualities onto other living things, but to phytomorphize ourselves—to stretch our own senses so far that we are able to sense an ever so faint recognition of what it is like to be them, within ourselves.

I stand by the water's edge, and I see her down there as her nose touches the ceiling of her known world, as it did when she first came up to swallow air to make herself float. It is a fluid skin where two oceans touch—the denser water-ocean below and the less compact and more luminous air-ocean above. Down there she is weightless, but her weightlessness is restricted to water. Even if she leapt, she'd be carried gently but urgently back inside. She would not soar easily into the thinner ocean above. Winged ones swim weightlessly up here, for their bodies are shaped to match the air-ocean's lighter density. Each body, it seems, is shaped for a particular regional gravity. She has her air bladder. Birds soar on wings. We two-leggeds dance on our hind legs. And trees stand perpetually upright in what seems like a lifelong embrace with gravity.

She drifts down there just below the seam of water and air, and she, too, can see all of us up here, flying, running, standing guard. I watch her there below, hardly moving, hardly breathing, simply looking up at me. I slip again from my own body and into hers, and I am looking up through the water. I observe the waves from beneath, I see the way light and shadows bop and frolic in odd angles and colors across the translucent ceiling of my water realm, and I look through the kaleidoscope above, and I wonder: What is it like to be living there, above? What other sentient styles are dwelling there, inside the air-ocean?

What is it like to have not fins but wings, to be clad not in scales that minimize friction and lubricate your movements, but in feathers that fan out and gather tiny air turbulences beneath your wings? Many winged ones look to the sunset to mark their direction and to fine-tune their magnetic sense. Like navigating underwater, navigating in air is a continuous creative play within a complex field of forces. What is it like to fly?[9]

Or what is it like to drink sunlight through a thousand light-responsive leaf-organs, each carefully tracking the sun as it moves across the sky? What is it like to experience the strange alchemy of your body transmuting those intangibles—sunlight and air—into something so substantial as wood? What must it be like to send part of your body foraging underground in search of water and nutrients, as roots do? What is it like to be permanently rooted in one place while coinhabiting a world of supermobile herbivores, sharp-mouthed plant-eaters whom you cannot outrun but whom you can confront through your own expressive style, sending chemical messages through vascular channels within and through the environing air, helping other body regions prepare for attack or attracting other creatures who would feast on the herbivores consuming your flesh? Perhaps here is a style of awareness far less centralized even than that of the winged ones, and far less centralized also than the salmon's decentralized, embodied awareness. Perhaps here is a style of awareness that expresses itself far more in a kind of immersed embodiment, or fluid exchange, or emergent conversation, between your wooden body and the larger surrounding bodies of the soil and the air. Perhaps this is an integrated mindfulness so vastly different from that of the salmon that you really must strain your own sensitivity very far to eavesdrop on them.

And yet, you struggle to try. Bit by bit, you disentangle the kinesthetic synergy of your own sensuous body, isolating and then stripping one sense after another. You slow your cold-blooded pulse to near-stillness; you quiet your gill breathing; you phase out magnetic geolocation; you let every recollection of light, color, of seeing, sink into the darkness below, for in this other body, in this trunk, you have never seen. As you intuit into the subtle, tectonic shifts within, you listen ever so cautiously for echoes that might spread.

Echoes.

Did you not just hear something?

Echoes that spread.

Recent research suggests that plants actually do hear, though perhaps not in the same way we humans do. Plants "benefit from some form of perception of substrate vibrations."[10] Soil, like the ocean, is a highly dense medium. There like here, wave energy moves rapidly and far, pregnant with signs, clues, meaning. The hairs that line the roots of plants can pick up the vibrations that move through this dense medium.[11] Hair in your own cold skin. Hairlike structures in the roots. A keen sensitivity even to the faintest vibrations in dense surroundings, and no physical need to evolve anatomical extravagances such as mammalian ears. What's more, this same research has shown that plants might actually have a way of producing sounds themselves![12] Plants actively send out signals. They might, oddly enough, even be described as "talkative."[13]

These faint echoes still roll through you, yet you let them pass, for your curiosity is already homing in on another question. What is it like to have undergone such a lengthy succession of adaptations that your side fins and your tail fin reshaped themselves into *feet*? And then, to have permanently risen from four feet onto two, spending your life balancing precariously between the land and the sky? What is it like to be a two-legged? You assume that just like trees, two-leggeds are parleying somehow both with the solid soil underfoot and the sky-dome above. You assume also that just like your own kin, they have some immediate rapport with the luminous sky-orb. For they, too, undertake a cyclical journey through the sky, not unlike the sun and not unlike yourself. Born from the ground, the two-leggeds set out to draw a circle across their lives, learning first to crawl on all fours and then to balance on their feet, migrating up into the sky as they grow, and then gradually shortening back toward the ground until, upon death, they return again under-ground to nourish the larger community of soil life. You are also quite sure that they also struggle to give voice to the felt reciprocity between themselves and the larger living land, and to the felt nourishment their imagination has been receiving from all these other breathing bodies since time immemorial.

How can you be so sure? Because as your own kinsfolk, salmon, have striven to make contact with the two-leggeds, they have responded. Twenty-three thousand years ago, they painted your image on the ceiling of a cave. Between then and now, their cunning hands and thinking heads have carved you in stone, cast you in poetry, and cast you even in the eloquence of their scientific prose. *Salmo salar* they call you—*the leaper*. Most recently their sciences, their rigorous methodologies, their patient empirical studies have helped them slowly and carefully piece together a story of salmon that is truly filled with wonder, even as it remains fragmentary and sketchy. It might be that in the face of all they are currently learning about you, they begin to realize that the most appropriate way to respond to your lives is to forsake the frivolous dream of mastery, and to surrender themselves instead to *thaumázein*, or awe.

Each of us has our own cognitive style, each our sensory skills, our uniqueness. And yet each of us is also listening to the discourse of other things. We listen, we embody what we hear, and we strive to respond. It is simply what we do. It is simply what we all do.

The Round River

Such an abundance of viewpoints, such a turbulent profusion of centers. Each folded into the storied, self-birthing Earth body in their own style and manner, and yet each as real and potent a form of experience as that of us, human animals. The sea—a realm full of sentience! Must we not learn to relate to it with the utmost care, forbearance, and respect, if we wish to enter it with our technologies and to receive sustenance from it? From within the watery expanse of this wet Earth, the notion that one land-mammal ought to be the center of it all appears selfish and naïve, and grossly inconsiderate to the larger-than-human community of life.

Aldo Leopold, in thinking of Earth, invoked the metaphor of the "round river."[14] His metaphor was born from intuition, years before scientists rediscovered the Earth as something utterly alive, and long before it became possible to speak in precise, scientific prose of the deeply entangled waves of action that roll on and on between the biota and the rocks, the air, and the water. Leopold did not know about cycles of participation that render the sharp separation between "life" and "nonliving environment" rather implausible. And yet he stumbled upon this beautiful metaphor, a simple image that succinctly captures what the Earth sciences have since been able to describe with increasing refinement: the recognition of a living sphere journeying through space, flowing, surging, and spilling round and round and round.

The Latin word *oceanus* has its root in what Homer called "the great river"—one river that spills continuously around the disc of the Earth. We might think of the great river of the ocean as a tributary to an even greater round river—the globe itself. Whales defecate to fertilize algae to seed clouds that are carried inland, to trickle through membranes, whereupon they flow on at the other side of the membranes as blood or sap or mountain stream or a moment of inspiration. The ground cracks open and becomes volcanoes, spewing out huge amounts of carbon dioxide and other gases where continental plates collide or drift apart, leaving the atmosphere recomposed. Continents birth themselves from the planet's molten interior and then sink back in again, lubricated by the liquid water that would have evaporated billions of years ago if this planet really were a *res extensa*.

And throughout it all, living creatures are folded deeply into the rolling waves of action, converting one style of vitality into another and contributing their share to keep the old round river adrift. Leafed ones metabolize the atmosphere and prevent it from coming to a standstill. Gillbreathers and lungbreathers metabolize the atmosphere all over again, offering portions of air-food back to the leafed ones. Fungi go about their inconspicuous subterranean networking and exchange information and nutrients across plant species, and break down tough organic substances that few others can return into the food web. An elephant matriarch labors to transmit a lifetime of knowledge to the next generation; a blackbird sings praise to the dawning day; a pod of killer whales leaps through boiling storm waves in a moment of pure bliss. I, a two-legged, take a deep breath, look around, and ponder the deep sense of wonder I feel at living inside this raucous, wild planet. All of us act to keep the waves of action moving that are particular to our kind. All of us labor to keep the boundaries open that weave us into the turbulence. We are all intertwined in these rolling waves as they move round and round the old body—composing it, sensing it, enriching it, meditating on it, celebrating it.

Somewhere inside these larger moving forces, the salmon attempt to keep their journey going. They stand guard by the semipermeable boundary particular to them, and they contribute their share to ensure that the round river flows on across their boundary. It is the boundary between ocean and land, between the very abundant hunting grounds of the northern waters and the far poorer, less productive grounds of continental watersheds. Each time they gather their refined fish senses to navigate back to the coast and find their way to good spawning grounds, they renew a contract between the seas and the land, a contract that extends into the memory of the land for as far as their unbroken lineage reaches. As they migrate up the river, those who stand by and witness their passage—nostrils twitching in recognition, mouths or maws or beaks watering, protein-starved bellies aching—understand and commit to the contract. It is a contract not of ownership but of relationship, a contract signed with blood and roe, with saliva and sweat and feces. It is a contract validated by pangs of hunger and the throbbing urge to reproduce, a contract ratified on the brink of starvation and in the midst of a community feast. Those who walk upon the land and those who swim up its rivers and those who forage inside its dark soils with their swarming roots, all pledge to give themselves wholly to the land, to take and to return. And the land pledges its complicity to all, to beaver, moose, owl, mallard, juniper, to salmon, and to us, the two-leggeds. It pledges to be food, shelter, nursery, deathbed. It pledges to be all of this at once.

This is the reciprocal exchange particular to the salmon. They are the keystone in a wide arch of other bodies. Their task is to keep this wave of action moving, to struggle to hold this specific boundary between the land and the sea semipermeable, to keep these many folks nourished.

To renounce modernity's monocentric structure is not to renounce all boundaries. Neither is it to say that centeredness is problematic. Boundaries are vital aspects of the larger unified Earth-river; they are locales of greater participation. They are unproblematic if they are kept semipermeable, if they are recognized as horizons rather than walls. Attempts to seal them are quick fixes both in philosophical terms and in physical terms, and they can hardly last. Metal fences or concrete tanks around the confined salmon, plastic lining atop their decaying bodies, or philosophical concepts that hold us and them categorically apart, all manifest a monocentric story that stifles mutually enriching exchanges across the species divide, and that firmly seals the boundaries between humans and the more-than-human world. It is surely folly to assume that the temporary explosion in food abundance, now heralded by the salmon industry, can last. Even as production continues to reach new peaks, we are starving ourselves—and the larger round river within which we dwell—in a far deeper and more acute sense.

Salmon nourish every aspect of our embodied minds and our mindful bodies. The nourishment we receive from them is fully integrated, wholesome. The Pacific Northwest writer Tom Jay has said, "The salmon is not merely a projection, a symbol of some inner process, it is rather the embodiment of the soul that nourishes us all."[15] They vitalize our flesh; they alert us to the eloquence of other-than-human ways of experience; they tutor us in the ethics of reciprocity. Through the instructions of the salmon, we are drawn outward and beyond the sphere of our strictly rational, strictly human-centered, egoistic concerns, beyond the permeable horizon of our thinking minds, beyond the permeable horizon of our species, and into the diverse, evolving, and constantly shifting vitality of Earth itself. It is a living Earth, a world full of multiple animate powers, all of whom vie with one another and interchange and sometimes struggle against one another.

We humans live in the depth of this interplay of different powers. Here, within, the salmon hold us to account for our actions. They remind us that as long as we continue to make gift-offerings to the more-than-human world, the round river of greater returns will not dry up. They remind us that as long as we nourish the old river with our attentiveness, our care, with songs and dance, careful science, thoughtful speech, and ultimately with our bodies, it will continue to nourish us in return. As Freeman House writes, "If it is salmon that chooses to lead some of us back to our immersion in the natural world, then our first order of business must be the survival of salmon, the health of the waters."[16]

If philosophy seeks to participate in shaping the conditions of human flourishing on an increasingly feverish planet, then it must see through the screen of self-centered concepts that hold us aloof from where we really, actually are. The ongoing challenge is to explore ways of thinking that articulate distinctions without making dichotomies. Philosophy can help cultivate techniques of slipping out of these received concepts and into this sensuous terrain, which we now know to be a living, dynamic, creative, evolving, nonhierarchical, multicentered, and fully integrated planetary presence full of other-than-human discourses, sentiences, storylines. The long habit of asserting our separation, and our central importance, is not only very depressing. It also causes a steady and irreversible torrent of species deaths, at a rate a thousand times higher than the common background extinction rate.

It is up to us to resist this burgeoning eclipse of amnesia. If the project of modernity is based in the double notion that humans are somehow of central concern and that the rational intellect rules supreme over the body, then modern cultures are likely to always have ecological crises, because such a tradition of thought disregards anything other than the human. It also overvalues that which can be abstracted, generalized, and quantified at the expense of qualitative ways of knowing. If instead we accept that we are embodied minds within an actively composed, storied, and more-than-human world, where sentience lives at large within the land, the air, and the oceans, and if we speak in ways that hold our mindful bodies open to each of these more-than-human sensibilities, we encourage a culture of steady and careful listening.

What is there to hear? An animate planet, a living Earth following its storied trajectory through deep time. Step inside and listen closely. It is the sound of the old river flowing round and round.

CHAPTER NINE

The Earth Ever Struggles to Be Heard

Sometimes your storyline is the only line you have to Earth.

SHARON DUBIAGO

*T*homas Aldwell was lying on his back, looking up at the sky. Beside him lay the bundle he had been hauling through the forest, consisting mainly of a very large can of coal oil. Off in the distance, the mighty Elwha River cascaded down a narrow gorge. The can of coal oil weighed Aldwell down. He had to make frequent stops. Decades later he would still remember this bundle. Each time he took a step forward, the bundle seemed to pull him another step backward.

Packing to his newly bought claim on the Elwha River was hard work. In the mid-1890s, there was still no wagon trail from the nearest town, Port Angeles. Nor was there a bridge. Other settlers rode horses, but Aldwell packed everything on his back. Sometimes he used a raft or canoe to cross a stream. When neither was at hand, he swam. But his hardships did not dishearten him. On the contrary, the primitive, aboriginal vitality and rugged beauty of the land grew on the man: "Those days packing into the Elwha cemented me to the Northwest," he would write in his autobiography half a century later. "The needle-covered Earth, the patterned tree branches, the sky, the fresh bracing woods' smell—all seemed to make me part of the Earth I rested on."[1]

Aldwell had not always lived here. Born and raised in the plains out east, he had come to the Olympic Peninsula to live the life of a pioneer.

"The West sounded easy," he later recalled in his autobiography. "'Land of Opportunity'; 'Plenty of Everything'; 'Generosity'; 'Friendliness'; 'A man's own worth and not his ancestors'!' These were catch phrases we all believed. Events proved them true."[2] Throughout his life, Aldwell would hold up these convictions as the torch that was to illuminate his pursuit of happiness.

Aldwell had arrived in Port Angeles aboard the *George E. Starr* just a few years earlier, and the town represented to him all he was seeking. It exuded the agitated pioneering spirit of a frontier town. It offered opportunity for those with the vision to create it. And it promised a great wealth of natural resources. The forests here had never been hewn; this was a community of stately giants, ready to be turned into pulp, boards, building material, affluence. A five-mile-long sand spit, created by sediment runoff from the Elwha River a few miles to the west, gave Port Angeles a natural harbor that could easily be turned into a center for industrial production and trade. Beyond the harbor, the waters of the Strait of Juan de Fuca overflowed with fish. Everywhere he looked, Aldwell found abundance.

Aldwell was a firm believer in the strength and resilience of community. He felt that the ability to include everyone's various talents, concerns, and voices enabled a community to thrive beyond its mere material standards. "There has always been a place for everyone;" he writes, "we could all be secure in the belief that we belonged."[3] He trusted in Divine guidance to safeguard him and his fellow men in their ventures. "It is my belief," he writes, "that our Supreme Being created our forests and their valuable crops of timber for the purpose of maintaining man as well as for the purpose of amusing him and providing for his recreation."[4] He also felt that material happiness alone was an insufficient foundation for a life lived happily and well. Once, during the 1940s, he was asked to give an ad hoc address to a Salvation Army convention, during which he expressed this conviction: "We businessmen think substantial and worthwhile things in this world are the business of material success which we may achieve. The Salvation Army people know that the substantial and worthwhile things in this world are the spiritual things and the good they can do to help their fellowmen. The day will come when we are all on our last resting bed and, looking up into the sky, we will then realize that they are right and we have been wrong."[5] A man of deeds rather than words, Aldwell nevertheless found himself drawn into the poetic presence of the Elwha when he encountered the river for the first time. He recollects that moment vividly: "My life had taken me to schools, to cities, to business, but suddenly that spring embodied all of life and beauty I thought I'd ever want."[6]

Little did he know that he would soon experience a powerful upheaval of perspective, a tremor in his perception that would leave him reconfigured, changed. He did not anticipate the power of a deeply held narrative to bring about fundamental change in the world. He also did not consider that sometimes it takes but a slight distortion of perspective, a minute deviation from the original angle, to set this power loose. In his case, it took a visitor.

Like Aldwell, R. M. Brayne was a businessman, and he came with a particular errand in mind. He was scouting for a place to build a water-power dam. And so he came to Aldwell's claim. The two men stood before the gorge in the stream, that narrow canyon where the water was channeled into a tight, roaring, foaming torrent. Aldwell was transfixed: "We looked at the canyon. Suddenly the Elwha was no longer a wild stream crashing down to the Strait; the Elwha was peace and power and civilization."[7] From then on, there was no turning back for him. "The property on the Elwha River always fascinated me but it was not until I saw it as a source for electric power for Port Angeles and the whole Olympic Peninsula that it magnetized all my energies."[8]

It was to become his life's work. He spent the next twelve years secretly acquiring the land. There were many setbacks. There were financial problems. There were delays. But he never lost track of his vision. His torch shone bright before him. The opportunities for both himself and for progress on the entire peninsula kept his vigor aflame.

The first dam was finally finished in 1912. But it blew out again the same year; contractors had failed to secure the construction in the bedrock. One person died in the flood, although nobody was later able to identify the deceased man. He was thought to be a traveler, a passerby. He was given an anonymous burial, and construction was resumed. In 1914, the Elwha Dam was finished. In 1927, a second dam, the Glines Canyon Dam, was built seven miles upriver from the Elwha Dam. They powered a pulp mill in Port Angeles, then the largest mill in the world. The Olympic Peninsula was ready for the boost in power. Its readiness echoed through the newspapers. "Development of Olympic Power First Big Enterprise to Start Lifeblood of the Peninsula," one headline noted.[9] In another article, from April 1911, printed under the headline "Putting the Elwha to Work," the *Sequim Press* noted that "nothing more helpful and desirable could be installed among us" than a dam that would allow for "the great power of the river" to be "converted from its waste and loss into a magnificent source of energy and strength."[10]

The dam had it all. Water—once wasted, flowing unharnessed—could be turned into measurable horsepower. Raw physical power could be channeled into a power for good. Energy and strength would be compelled to facilitate settlement, cultivation, improvement. The Elwha would never stop cascading down the Olympics. This turbulent spring of goodness would never cease. All could be secure in the belief that they all belonged. Everything was well.

Or was it?

Let's rewind and discover the Elwha River one more time.

⁓

The river originates high up in the alpine reaches, where glacial meltwater trickles down lichen-covered rock. Time up here is measured by the pace of the mountain eroding away beneath this trickle. Drop by drop, puddles form. Dripping strings of water follow gravity valleyward. They form rivulets, then streams. The mountain is in motion. It moves with the pace of the seasons, with the pace of rock dust, with the pace of water. It is as it is. It is as it always has been.

The river drops quickly through temperate forests, gathering momentum and volume as it is fed by tributaries before flowing out into the strait. These forests, like the mountain, have their own pace. Since the glaciers of the last ice age melted away, the forests have seen repeated cycles of growth, adolescence, maturity, calamity, and rebirth. Long cycles of birth, death, and rebirth appear to have swept down and across these slopes, each clothing the mountains in hues of maple extravagance and evergreen. Now, as ever, the river runs through it all. Those who live by the river, and those who live in it, know that it is their lifeblood. It is the vein that connects and nourishes all. Black bears, raccoons, otters, elk, bald eagles, vultures, and cougars are some of the more conspicuous creatures. There are hundreds of others.

There is one family of creatures in particular. They, too, are children of the river. But they are also travelers, vagabonds who will venture out to sea when their time has come. They will leave. And then they will return. When it is time for them to give birth and to die, they will find their way back home. Hundreds of thousands of them will find their way back. They are cutthroat, char, steelhead, Chinook, coho, pink, chum, and sockeye. They are trout, and they are salmon.

There is also another family. They are known up and down the coast as Klallam, or the Strong People. The story of their name tells of a people not only strong in body, but strong also in character, a people resourceful, and cunning. They did not choose this name; they received it from others. It was a token of respect.[11] The Klallam know that they, too, are creatures of the river. Their elders have passed down the knowledge. An unbroken line of teaching reaches back into time immemorial.

In the beginning, they, too, stepped out of this river. They can walk you to the precise place by the river where the Maker shaped them from riverbank mud. Beside the perpetually cold glacial current, at a site known in the Klallam language as the place of "coiled baskets," lies a big, flat rock. Inside the rock are two deep hollows that are shaped like coiled baskets. The Klallam know that it was from these holes that the Creator scooped up dirt. And out of this dirt he then made the Klallam. He bathed them, and he blessed them, and then he sent them on their way. They went out, and, like the salmon before them, they have never failed to return. They often come traveling upriver to this place on vision quests. They bathe in the Elwha's current, and they prepare themselves for encounters with their spirit powers. The place of coiled baskets gives them guidance for their lives. They will let a hand slip inside one of the hollows that is filled with water, and they will pull out the first thing their cold fingers touch. Then they ponder the thing they have pulled from the water, whether it is deer hair, feather, or leaf. It symbolizes what the future holds in store for them.[12]

The Klallam are a community of several bands and tribes, who all belong to the larger community of Coast Salish people. The Salish extend eastward into the far reaches of the sound, westward to the rain forest coast, and northward across the strait to shores on the other side. To the south, mountains form the horizon of their habitation. Like all Salish people, the Klallam, too, inhabit a mindscape richly layered with story. No one knows how many strands of story there are. Some tales spring to life and then slip again into oblivion. Others are carried on but metamorphose along the way. The most resilient strands are carried through the generations with few or no changes. You never change a story, the elders will say.

In the Salish universe, at the beginning, the Maker is lonely, and so he arrives at the idea to make the world. He conceives a song and sings the world into creation. The Maker gives life to the world, and he decides he wants to have as many voices as possible. All the living beings he makes are given different forms, and every one of them is given its own voice. The Salish people share this intuition with other humans across the continent: the notion that in the beginning, all the animal people speak the same language. All can understand one another clearly. All are given the gift of voice so that they can participate in singing the world.

The creatures receive another gift from the Maker. They are given immortality. In the beginning, nobody is hungry. Nobody is cold. Nobody has any needs. Nobody ever gets sick. Life is perfect.

The elders say that perfection was precisely what was wrong with the first world.

In the perfect world that the Maker has sung into existence, nobody ever needs each other. In this world without time, where nothing ever changes, and everything is eternal, the first people are like children. They never grow up. Eventually, they cannot bear this any longer. They start sensing that what they want more than anything, is to have a need. They want anything that will let them perceive change. An upheaval, or a rupture. And it is impossible to find.

At long last they begin comparing themselves to the others. "I have longer teeth that you do. I'm better."—"I have keener eyes than you. I'm better."—"Look how beautiful my feathers are."—"Look at how strong my tailfin is." Lacking any meaning in their lives, they become self-righteous, wicked, malicious. Driven by the craving to distinguish themselves from everyone else, they become narcissists. They begin living at the expense of everyone else. This, the Salish stories say, is the original sin of the world.

Eventually, the betrayals, the belittling, the abuse, the monstrosity, and the corruption in the world are so great that the Maker recognizes it cannot go on. "I must do something. I must come to the aid of the world I have sung into being."

The Maker sends a helper, henceforth known as the Son of the Creator. The Son of the Creator is given the task to pass judgment on the world. The Son of the Creator comes in different characters. In some places, he is raven. In other places, he is coyote. Here, on the outermost edge of the continent, the Son of the Creator haunts the stories in disguise. Nobody knows who he is. Nobody ever recognizes him unless he chooses to reveal himself.

At first, the Son of the Creator wanders about passing judgment on the misdeeds of individual creatures. But there comes a moment when that proves insufficient to heal the wrongs of the world and to alleviate the overbearing suffering. He must take his judgment one step further. He knows that it is time to strip all the beings in the world of their immortality.

The first creature to meet her death is a young girl. For many years, this wretched girl has suffered abuse from her cruel mother. She has received too many blows, heard too many harsh words, been belittled too often. She can take it no more. The Son of the Creator goes to her and gives her the choice to go with him. She considers the choice she has been given. Then, she takes his hand, and they walk away together. This is how she dies.

From then on, death spreads through the world. It is the final judgment on all the creatures. Everyone is held to account for how they have treated the world. Everyone is held to account for having become corrupted in their hearts.

The suffering of the world is not yet mended. If anything, the creatures are even more miserable now than they were before. They are no longer immortal, but they are also cold, and they are hungry. Many of them are hungry for the first time in their lives. They do not know what to make of this unpleasant sensation. In this new world they inhabit, the world of the mortals, they all suffer. They find no solace. They do not know where they belong. They are lost.

Once again, the Maker looks upon the world he has created. He realizes that there is too much suffering in it. He will come to the aid of the world one more time.

The Maker goes to the spirit world to speak with the first ancestors. They are the ones who have moved on. They have paid their dues, and they have regained their immortality. They are out of the pain and out of the suffering. He goes to them, and he says, "Your children and your grandchildren are suffering terribly. Would any of you be willing to go back into the world and to feed them?"

Some of them will not go back, but there are some who will. Some choose to return to the world, to ease everyone else's suffering. They come to give their lives so that others might live happily and well.

They are the first salmon.

And so, a bargain is made between the salmon and the other creatures. The salmon will swim up the rivers and feed their children, year after year after year. In return, their children will honor and respect them. They will put the bones of their ancestors back into the water. They will remember the teachings. They will respect the teachings, and they will uphold them for the next generation to learn. They will make sure that the line of teaching is never interrupted. In return, they will live well and in perpetual abundance. The original agreement between the Coast Salish people and the salmon is clear and simple: The salmon will come back to relieve the suffering of their children. But if the children cease to honor them, the ancestors will stop returning.

Of course, this does not happen. The Coast Salish remember, and they uphold the agreement. This is the way it is. It is the way it always has been.[13]

Narrative Hegemony

Having lived for millennia on the precarious edge of the continent, the Lower Elwha Klallam tribe belonged to that complex network of Coast Salish people who shared an intimate animistic participation with the land, the air, and the ocean.[14] Language, they felt, originated with the wind. Thought resonated with the articulateness of the ocean, of volcanoes, of ancient forests. And it resonated with the eloquence of the salmon. Salmon were their oldest ancestors, their most highly revered elders. A richly textured familiarity between humans and salmon was part of their genesis stories and found articulation in a Klallam land ethic whose core principle had always been an "original agreement" with salmon.

As settlers arrived on the Olympic Peninsula throughout the seventeenth, eighteenth, and nineteenth centuries, a new story began inundating the land. That new story suggested that there were no voices inside the land worth listening to, no powers that should interrupt the perceived improvement of the West. Forests, mountains, and rivers were cast into a new light, as were those who dwelled there, including the original humans and the salmon.[15] Before long, the Elwha River and the old-growth forests surrounding it were turned into "organic machines,"[16] compelled to yield raw power and raw material, "resources" that were to bring "progress" to the West. The original humans and the salmon were nearly entirely overlooked. Thomas Aldwell's three-hundred-page autobiography, *Conquering the Last Frontier*, mentions the Klallam people but once. Peace, power, and civilization: Those were the magical words that would convert the Elwha "from its waste and loss into a magnificent source of energy and strength."

Reinterpreting the Elwha River community as an organic machine also brought about enormous suffering. Thomas Aldwell's two dams on the Elwha were built without fish ladders. Returning salmon crushed their heads in futile attempts to overcome a concrete wall that could not be surmounted. The number of returning salmon dwindled quickly from four hundred thousand to less than four thousand. With the salmon disappearing, the Klallam were cast into a struggle for survival that became physically and culturally desperate. The near-total loss of the salmon meant the loss of an economic base that had nourished them since time immemorial. With their oldest ancestor suffering so greatly, the Klallam strained to uphold their identity. Substance abuse, domestic violence, unemployment, suicide, a loss of self-esteem, and a loss of language each took their toll on a people whose economic and cultural heart, the salmon, had nearly entirely stopped beating.[17]

Since its beginnings, modernity has advocated a kind of "narrative hegemony," the expectation that one "right" story would emerge by virtue of technological and moral progress. The modern story was fueled by

epistemological optimism, or the idea that as the rational intellect withdrew from the world and cut ties with tradition, it would gradually come to
a clearer and more perfect understanding of the world and of humans' place
within it. The mind-body split and the human-world split would eventually
overcome all uncertainties and create a world with no more imperfections.[18]

In Descartes's age, the un-centering of Earth had left Europe's
educated citizenry in a precarious state of distrust. In that context, the
Cartesian split became a remarkable success story. Descartes responded
to his contemporaries' widespread insecurity by offering a desirable
counternarrative. The Cartesian split was a victory of the rational intellect over the body, a victory over convention, boosting a general mood
of epistemological optimism. Every new achievement of the rational
intellect would bring Europeans a little closer to what Bacon had so cunningly spoken of as the "millenarian promise of restored perfection."[19]
The new story quickly gained momentum: The rational intellect would
look ahead and not turn back. It would create an immaculate world, a
world dictated by reason alone, no longer vulnerable to outside forces.

But the story of humanity-as-separation not only alienates mind
from body, and it not only estranges humans from river deltas, mountain
plateaus, aspen groves, bears, or salmon. It also splits "our story" from
"other stories." It reflects an ambition to "otherize" that which is perceived to be "outside." This creates a strong pressure to deny that there
might be other narratives, other stories that can function as primary
sources of intelligibility and value.

And so the Cartesian split also tends to marginalize other-than-modern
(typically oral and indigenous) cultures. There is a parallel logic at work
here. Oral, indigenous cultures who have not endorsed the mind-body
split, or the human-world split, might be perceived as "not quite human."
They might be perceived as "more animal than human," as having "not
yet" adopted modern ways, and as somehow "lagging behind." The modern split from the world elevates itself to a historical norm by which
all are evaluated. Just as the perceptual othering of salmon can make
it seem normal, even desirable, to exploit the salmon, the othering of
nonmodern humans can make it seem normal, even desirable, to exploit
those humans who have not subscribed to the dominant narrative.

Music That Rises from the Earth

The modern experience of time is a peculiar thing. As Freya Mathews has observed, modernity's hallmark is radical change, a commitment to the ever-emerging new, a deep dissatisfaction with the given. Modernity fetishizes a radical discontinuity with the past and dissociation from tradition. When our attention is directed only to the new, time is perceived as moving in a rectilinear way, a straight line, from a distant past toward a not-too-distant future. This sense of time has trouble turning back and is more skilled at seeing what lies ahead. This linear time creates a narrow horizon of attention.

From within the narrow temporal horizon of the dominant narrative, it becomes arduous to attune to other voices in the land. Think of the longevity of a redwood forest, or the birth of mountains, or the primordial dance of continental drift. Inside an attention span that perceives only the succession of the ever-emergent new, the world is so much less alive. It is so much more inert. Drifting along a line of ever-novel aspiration, never settling into the truly vast and unfathomable expanse of the presence, moderns could come to think that there are no other storylines in the land or in the sea, no voices to parley with but their own.

———

Until contact, the Salish people see no need to count the passage of years, for in their world there is no point zero, no singular event that defines the topography of time. They see no need to equate the thick and complex dynamism of the sensuous terrain, the land's storied unfolding, with what the new humans call history.[20] They live inside a thick present, enriched with countless strata of memory. They have been in place so long that they have developed an intimate, storied experience of the partially structured and partially anarchic vibrancy of that place.[21] They know that their watershed is anything but inert. It is pulsating with life. To spontaneously experience all things as alive, as the Klallam do, is to be gradually drawn into the thick expansiveness of the present moment. It directs the attention not so much to what is new, but what is now. The animistic experience draws the observer into a layered, diverse, complex, creative, active, deeper and denser experience of the present.

Everything here has a voice. And the Klallam know that their culture originated in the midst of this vibrancy, in the patterned emergence of this Earth itself. Their stories speak with the voices of the land. In fact, the Klallam language itself has become "a language full of the sounds of water, of the river."[22] With persistence, and alertness, the Klallam have slowly learned to translate the voices that emanate from their entanglement with the present into human dialect. The Klallam seem to receive their language from the texture of the living land itself, from gusts and blasts and howls, from splashes and spray and splatter, from growls and screeches and barks. And as they eavesdrop on the conversing of the wind in ancient cedar stands, and as they attune their senses to the pattern of snowfall, and as they become more receptive to the songs of killer whales, they begin to lend their own voices to the land in return, to nourish those who have given them not only sustenance but a voice. The Klallam return to the land that particularly human nourishment that is memory, held in story. The exchange between the Klallam and the living land is reciprocal, the conversation dialectic. Dwelling, to them, means becoming participants in an embedded and creative act, enriched by a habitual inclination to listen to the many other voices inside the land. It means inhabiting an articulate landscape in which the gift of voice is being continuously received and offered back in return.

During a visit to the Elwha in 2012, I met Andrew Fitzgerald, a young Klallam tribal member who has been an eloquent campaigner in the indigenous renaissance movement.[23] He had suggested that the two of us meet atop the sand cliffs of Dungeness Spit, a large sand spit that extends into the ocean some twenty miles east of the Elwha. A thick haze was creeping in from the sound. The distant shores of the Canadian border in the north were being swallowed by the haze. The sun stood low on the western horizon. A cold and quiet breeze was moving through the vegetation around us. As we spoke, he became aware of the breeze. And then he said to me:

> The reason that we speak the way we do—the reason that we sing songs in a beat way—is because we're out here, and the songs have to travel around the trees. And that's the reason that they're

deep. It's also because of the way the ocean roars. When in our songs we go like, OOOHH, it's because when the ocean comes in, OOOOOOOHHHHH, like that, or when you hear the river, OOOOOOOHHHHH—those are the sounds we hear. Those are the sounds the ocean teaches us, and the river teaches us. That is why the songs begin the way they do. Those songs, they get told, and they get dictated by the way the wind travels through. . . . . All this language in the world, that's the wind. That's what our people say. The most powerful language in the world is the wind. Some of our songs don't have words because the most powerful—the wind—can *change* the whole surrounding area. The wind can change that. And because of that, it is the most powerful. And so our most powerful songs are the ones that have no words at all, because wind precedes . . . the human interpretation of the world. It can change and it can constantly be . . . made different. But it doesn't make it wrong.

Before contact with the Westerners, such an experience seems to be easily palpable for the Klallam community at large. It resonates through the voice of every woman who addresses her audience in oratory, and every man who chants at a gathering, and every child who fumbles for the vowels and rhythms and intonations of its mother tongue. The Klallam hear each other speak, and they hear the land speak to them and through them.[24]

When there is no point zero by which to mark a calendar, no single historical origin that fixes all subsequent events in sequential succession, no uniform flow, the attention of the embodied mind turns toward the textured, patterned unfolding of the *presence*. The sense of time that shapes itself in response is finely integrated into place. The rhythmic patterning by which things happen here on the Elwha is specific to this place and incomparable with any other.

The majestic Olympic Mountains prevent the Pacific rains from marching in from the west, making this northern edge much dryer than the eternally drenched rain forest coast.[25] Dense, cold fog can suddenly gather in the waters to the north, shielding the large island far off in the distance and sending its frigid, cloudy breath up the banks, even at the height of summer. Some of the coniferous trees here have massively thick, tall trunks, and they are so old that their life span is practically incomprehensible for the human animal. Cold and hot summer airs rub passionately against one another until they erupt in enormous thunderstorms, cloudbursts that scorch the sky from morning to night, sending bolt upon bolt of jagged fire into the ground, making some of those wooden ancients crack like bones. And every few human generations or so, certain pyramid-shaped, solitary mountain giants just barely visible in the direction of the rising sun might erupt in a massive explosion of lava and ash, darkening the skies, making the earth itself quake and rumble.

This is all part of the felt temporal quality here inside the Elwha watershed: the sharp *Zrrrkkkk!* of lightning splitting the sky, the cold and obscure breath of the ocean fog, the dormant ferocity of those volcanoes in the east. Each thing imparts its subtle influence on the overall experience of this place's time. Like any other place, so, too, the Elwha speaks uniquely. It articulates its own time.

But there is also a temporal eloquence here that is shared by many—human and other-than-human—up and down the coast. In fact, it is the temporal lingua franca along the entire Pacific Rim: From California in the south to Alaska in the north, there is that annual return of the five salmon nations, the Chinook, coho, pink, chum, and sockeye. More than any other pattern or pace, these five nations' cyclical journeys give temporal coherence to the region and imbue so much of the Pacific West with a felt experience of time as not only layered and specific, but also, importantly, round.

The Klallam are a wealthy people, and they know that they owe much of their wealth directly to the salmon. Living within round time—within salmon time—their awareness is drawn again and again into the thick of the presence. Year after year salmon bestow upon the watershed the gift of their astonishing abundance, signaling a structured renewal, an ongoing creativity of which all might partake.

All five Pacific salmon as well as numerous steelhead and anadromous cutthroat inhabit the Elwha's cold, rapid glacial currents. Their return to their Elwha birthplace every year is heralded by the ten discrete seasonal salmon runs that populate the river. Chinooks, or spring salmon, arrive from April through July. Sockeye and chum arrive in July. Pinks, or humpbacks, are harvested from August to October. Coho, or silver salmon, arrive from October through December.

The humans in this watershed recognize the salmon as their cultural and economic bedrock.[26] The Klallam pursue the salmon using lines, gillnets, reef nets, spears, and traps. Fish weirs tend to be the most productive method for fishing. The Klallam set up their weirs in estuaries, rivers, and smaller streams, using fir branches, vine maple saplings, and cedar limbs to construct the poles and the webbing. Weirs are built so that they can be left in the river throughout the year. They take a community effort to build, and they benefit the entire community. When the salmon arrive and the weirs block their passage upstream or guide them into traps, the entire community shares in the catch. These weirs would be effective enough to block entire rivers, if the Klallam intended to. But weirs are always constructed with built-in escapements, allowing some of the salmon to pass through uncaught.

Each year when the first salmon start pushing up the river, the Klallam are there awaiting them. Songs intertwine with the mist on the river. Voices of everyone from elders to toddlers incant the familiar greetings. The lead salmon is considered the chief of the salmon people. He is cooked, cut, and gifted to the tribal elders in a ceremony led by a shaman. His head and bones are arranged with great care on a cedar mat. The raft is then placed carefully into the river's currents, which will return the chief's remains to his people, who live in houses below the sea. He will tell his fellow travelers how honorably he has been treated, and that it is alright for them all to move up the river.

The First Salmon Ceremony is one among many first fruit ceremonies.[27] Like the other first fruit ceremonies, the First Salmon Ceremony gives structure to the perceived roundness of time. It instills in the people a sense of gratitude, a sense of alertness, and the knowledge of the origin of their provisions and their wealth. It helps the people remember the original agreement they have with their first ancestors. Stories explore the theme of the original agreement in unceasing variations. The practical knowledge and moral wisdom held in stories then trickle down into every other aspect of their lives.

———

Beginning with Euclidean geometry, our own narrative tradition has thought of time and space as fundamental and separate aspects of the phenomenal world.[28] Space is thought of as absolute and static: All is basically there; it just *is*. And time is a separate dimension, a steady and even flow that bisects space perpendicularly, always going somewhere. Time is thought of as an arrow, and it resists any resonance or integration with the world's spatial dimension.

But to the cautious observer, no landscape or region is wholly inactive. Each place has a unique style of expressing its innate tendency for transformation. There are the sudden and unpredictable bolts and rumbles and quakes and explosions in the sky and in the ground. There are the tidal waves, ageless, tireless, dependable. There is the incredible slowness as the West Coast conifers grow up into the reaches of the sky. There is the patterning of flower bloom, or the seasonal flux of the night sky.

Through an animistic experience, the perception of spatiality and the perception of temporality seem to compose one another dynamically; they seem to be but different aspects of the experience of being an embodied participant *in place*. Where the disembodied mind sees "objects" passing "through time," the place-sensitive, embodied observer witnesses events unfolding from within the depth of her own immersion in place.

What if what we are used to calling the progression of time might be more clearly understood as a measure of the world's creative emergence, as a measure of each thing's immanent creativity? To experience oneself *in time* would be to converse within Earth's continuous creativity. Time would be neither transcendent nor linear, nor would it flow at an even rate. It would rather be immanent in the richly textured land itself.

And what if what is commonly spoken of as "time" is actually the land's voices, its music? A music that is fully of the Earth body. Time might simply be the structured quality of the land singing itself into being.

In the beginning, the Maker conceives a song. He sings the world into creation. He wants as many voices as possible. And so every creature is given their own voice. All are given the gift of voice so that they can participate in singing the world. Perhaps then, this expression, "in the beginning," does not accurately refer to a singular, historical event. Perhaps the expression rather tries to capture the fleeting, bodily experience of an ongoing creative thrust: music that does not cease to rise from the earth. Voices that pour themselves into the fugitive present.

Creation

These differing perceptions of time also impact the way both narrative traditions understand their creation myths. Such myths are a concise articulation of a culture's relationship with the felt experience of the world's creativity, or creation. Consider the following encounter between new and original humans, told by the Sioux doctor Dr. Charles Eastman:

> A missionary once undertook to instruct a group of Indians in the truths of his holy religion. He told them of the creation of the Earth in six days, and of the fall of our first parents by eating an apple. The courteous savages listened attentively, and, after thanking him, one related in his turn a very ancient tradition concerning the origin of maize. But the missionary plainly showed his disgust and disbelief, indignantly saying: "What I delivered to you were sacred truths, but this that you tell me is mere fable and falsehood!" "My Brother," gravely replied the offended Indian, "it seems that you have not been well groomed in the rules of civility. You saw that we, who practice these rules, believed your stories; why, then, do you refuse to credit ours?"[29]

From the point of view of that particular linear historical narrative, there is a categorical distinction between sacred truths on the one hand, and mere fable and falsehood on the other. Surely, modernity has inherited this sense of historical time from its Christian roots. Time, in that view, proceeds forward with linear directedness, moving through history like an arrow. As such, the moment of revelation is thought to be something that can be localized accurately on history's arrow. It becomes quite inconceivable that there could be a very different understanding of revelation, one which needs not invoke the concept of history. That other sense of revelation locates itself, once again, in place. As the historian Vine Deloria observes: "The places where revelations were experienced were remembered and set aside as locations where, through rituals and ceremonials, the people could once again communicate with the spirits. . . . [R]evelation was seen as a continuous process of adjustment to the natural surroundings and not as a specific message valid for all times and places."[30]

From a historical perspective, the notion of creation implies a singularity in time, an event effected and completed in historical time. From a place-based perspective, that which we commonly call creation is rather experienced as an ongoing creativity, or, in other words, as an expansive activity or agency inherent in all things. "At no point does any tribal religion insist that its particular version of the creation is an absolute historical recording of the creation event or that the story necessarily leads to conclusions about humankind's good or evil nature," Deloria observes. "At best the tribal stories recount how the people experience the creative process which continues today."[31]

In the account of the Sioux physician Dr. Charles Eastman above, the new humans were quite unable to understand that the indigenous were speaking factually about the origin of maize, just as they experienced it. The new humans had not themselves been prepared for the way in which each landscape can simmer with its particular creativity, with its own swarm of voices that strive to make themselves heard. They were rather ill-prepared for the possibility that several true accounts might be able to exist side by side of one another (each being true in, and responsive to, a particular place). Through their historical and placeless concept of creation, the new humans felt a divine calling to bring others under the auspices of what they perceived to be the only true way of conceptualizing creation.

In a place-based conception of the land's creativity, on the other hand, it is rather incomprehensible that one would experience a calling to missionize, to spread a gospel. Such a calling would run counter to the experience that each place expresses its unique creativity. If revelation is inseparable from the metamorphic topography of places, then it must be quite perplexing indeed to encounter visitors who speak of "eternal truths" that would be equally true everywhere. And to uproot revelation from place, to try planting it elsewhere, would invite for the possibility of suffering, as it could not be done without displacing someone else's experience.

———

Vine Deloria observes that throughout pre-contact North America there is a widespread appearance of emergence myths from the underground. "The Navajo legends begin with an account of the emergence of the Navajos or First People from the underworlds," writes Deloria. Similar stories are known to the Pawnee, Arikara, Pueblo, Mandan, and several other tribes, all of whom have variations on this theme of "emerging from the underground," prompting Deloria to speculate that "some common experience must be shared by some of the tribes."[32]

Perhaps at least part of that common experience that Deloria speaks of is precisely the experience of a felt immanence, of being birthed inside a living Earth that continually births itself into the patterned, storied, stratified, and metamorphic present. Perhaps at least part of the reason for why emergence myths are relatively common throughout the continent is that many of the indigenous cultures practice an animistic participation with the land, the experience that the land continually sings itself into being and shapes itself into the mindful body of coyote, or raven, or salmon, or human, to linger there for a while until each sentient body returns again, in the end, to nourish the land's wider and more mysterious creativity.

Along the Pacific Rim, the most articulate and distinguished incarnation of this creative process that continues today might actually be salmon. This might be what the Salish origin story of the first salmon poetically captures. The salmon's return from the spirit world might signify the very principle of the world's self-emergence into unconcealment, its self-delivery right here into the presence. Originality might be throbbing in every grain of gravel, in every piece of driftwood on the beach, in every play of shadows beneath the trees, but it might be that this primordial impulse simply expresses itself far more lucidly through some things than through others. And for storytelling animals who live along the edge of the Pacific Ocean, it might find its soundest resonance through salmon.

Further inland, the Navajo culture knows of a character who, in striking parallel to the salmon out West, also signifies the land's immanent creativity, this wellspring of all phenomena, this wombish quality intrinsic to the very experience of Being. In Navajo country she is called Changing Woman. Changing Woman is the most venerated of all the Navajo's Holy People. Like the salmon, so, too, Changing Woman repeatedly ages and rejuvenates in a seasonal cycle, and she is considered the mother of all Navajo people.[33]

Perhaps both Changing Woman and the salmon are held in such high esteem precisely because they are such luminous embodiments of the experience that the world is birthing itself. Through the salmon, as through Changing Woman, storytelling animals might be able to express in words this rather ineffable creativity. Through stories of salmon, or of Changing Woman, humans might be able to bring their embodied thinking into resonance with a mystery that cannot be fully owned, explained, or even thought: It is motherhood; it is the ancestral wheel; it is everywhere and at the same time wholly in each tangible thing; it is simultaneously ancient and of the present; it is ever expectant with possibility, ever wheeling itself back into the vitality of the presence. It is also not transcendent but fully immanent in the phenomenal world.

And perhaps that which the salmon and Changing Woman signify and embody cannot be spoken of any more directly than through allegory or story. There might not be an image or phrase or thought that would be able to stand fac-to-face with this enigma, and to see it all at once. Perhaps it is rather like the sun: Attempt to look at her directly and you will not be able to see her.[34] And yet you can indeed look the sun directly in the eye. What it takes is a subtle shift in perspective: The sun pulsates in each wheezing grunt of elk high up in alpine pastures. She warms the blood of each predator seeking stealth beneath the tree elders in the valleys. She comes to sentience through each sentient creature in the ocean. And she is in each word spoken in praise of her return in the spring. She, too, signifies motherhood, a wombish quality, a nurturing that is unreserved and unconditional. She, too, signifies and embodies the ancestral wheel. Each thing here inside Earth's swirling sphere of life is her, reincarnated, reembodied. She is not only "up there" but also fully immanent "down here," and it is right here inside her immanence that we are able to encounter her. We look into each other's radiant eyes, and we see: Each of us is sun.

Likewise, one might not be able to see this unifying creativity directly, except in the way it expresses itself through each thing. It slips into unconcealment at the same time as it slips back again into concealment. It never simply is. It does not rest long enough for any observer to confine it through a word or thought or image. It is not the static "nature" of modern metaphysics. It cannot be fully thought because thought itself is of it, inside of it. Thought cannot fully step outside this mystery.

But this mystery is not transcendent; it is palpable. It cannot be captured by thought, but it can be experienced by paying attention even to the minutest rousing inside the land. Something ever stirs. Someone ever urges herself into the midst of the awareness. There is a dual quality to this animation, this mystery, and yet it is not dualistic. It is the creative unity before and beyond all forms, *and* it is these forms.

In this palpable, immanent mystery, we find an experiential knowledge of the sacred that sees no need to abstract the divine from the world. The divine is rather seen as that which delivers itself from the enduring well of the possible, to become flesh inside the world, and then, once it has consumed itself as flesh, to dilute again into the well. It is a sense of the sacred that has no need for a transcendental realm; instead, it embraces a radical immanence: The sacred expresses itself here inside this world, inside this atmosphere, inside this flesh.

This is precisely the responsibility the salmon choose to take upon themselves when they alone come forward to return from the spirit world. Their gift is the coming into flesh as food, and when they do, they draw all others into the wheel. Through their self-sacrifice, the very act of eating and being eaten, rightly practiced, becomes a step into the sacred. Fritjof Capra has fittingly observed that "sacrifice" in the original sense means "making sacred."[35]

And surely, the salmon's return to initiate this act of self-sacrifice is not a singular historical event. Their return from the spirit world, their invitation to step into the sacred, is a recurrent act of revelation, and of creation—one that cannot be so well understood as a singular historical event. It is rather a poetic quality inside the storied land itself.

The Moment When History Began

When European settlers came to the Pacific West, these differing conceptions of time and place came into contact with one another for the first time—the arrow of history pierced the felt roundness of time. The first time the arrow struck was in the year known to the new humans as 1592, the year when new humans first came sailing along the unmapped Pacific Coast.[36] Most of their names are now forgotten. But one name, that of their leader, remains in living memory. He was Juan de Fuca.

Born in Greece in 1536, and sailing in the service of King Philip II of Spain, Juan de Fuca had journeyed to the Northwest to find the Pacific gate to the legendary Northwest Passage. He knew that if he found it, his rewards would be beyond any earthly goods. Once his name was written onto maps and into legend, he would be elevated to the status of immortality. If he succeeded, he would be beyond the common fate of mortal men, which is forgetfulness.

There is some controversy as to whether it was Juan de Fuca himself who named the Strait of Juan de Fuca after himself,[37] or whether the British fur trader Charles Barkley later named it after its first European explorer in 1788.[38] In any case, the Greek who sailed under the Spanish flag in 1592 succeeded. He passed through, never having set foot on either side of the strait. There was no need. Maps became his life's memorial.

Nearly everything in Juan de Fuca's travel account was erroneous. The strait he sailed through was not the entrance to the Northwest Passage, though he reported that it had been. The Strait of Juan de Fuca leads into Puget Sound or, as it is also known, the Salish Sea, where it fizzles out among the foothills of the volcanic Cascade Range. De Fuca was also wrong about the latitude, claiming that the entrance to the strait was at 47° north latitude when actually it was at 48°. Furthermore, the description he gave of sailing the strait bore little resemblance with the land itself. Finally, he was wrong about another conspicuous detail: In his accounts, these lands showed no sign of human presence. They were a no-man's-land, a tabula rasa, vast and beautiful but empty. Like Aldwell later would be, de Fuca managed to be entirely blind to the presence of other humans. From de Fuca's perspective, this was a land devoid of people, devoid of names, devoid of stories. It was an inanimate land, a place that did not speak. Being empty and inert, it was there for the taking. It was an empty book into which Europeans could write their own narrative. Everyone else was on the ontological margins. Everyone else was nonexistent.

Housekeeping inside Time's Roundness

Meanwhile, the Klallam and the other original peoples go on thinking of the Salish Sea and the lands around it as their home. They go on trading, fighting feuds, visiting relatives, dancing, addressing audiences in oratory, arranging potlatches, giving thanks to their ancestors. They go on nourishing the land with rich layers of lived memory and cultivating the ground with story. All this unfolds within time's felt roundness. As the seasons pass, they go on structuring their economy through an annual hunting and gathering cycle that follows the migration, maturing, and growth patterns of the salmon, as well as many of the other plants and fellow animals they pursue.

The arrival of spring marks the beginning of activity. Throughout spring, summer, and fall, all Klallam are engaged in gathering, tending to plants, hunting, and, of course, fishing. The provisions they do not consume right away are dried and preserved for the leaner season.[39]

By late summer, the Klallam habitually prepare for a longer canoe journey. Many of the roughly three thousand people from some eighteen villages along the beaches will embark on a trip east, through the strait, and then south into the needle-shaped canal that cuts deep into the mainland.[40] Among the provisions they bring are dried clams and fish, for the Klallam are famous for their clams, and they know that the clams are sought after and prized in the Skokomish settlements along the canal. Many of them will spend August, September, October, and even November and December away from home, fishing for dog salmon, trading with the people of these other shores, and reinforcing kinship ties. Come winter, most of them will steer their canoes back home. December and January are traditionally a time for resting, and for enjoying time spent together in the villages. During those lean months, the land bestows its gifts more modestly. Still, the Klallam know they are wealthy. They have stacked enough cedar boxes with dried berries, herbs, game, and salmon, and there is no hunger. Winter is the time for ceremonial renewal, for singing, dancing, visiting. It is also the time for holding one of the famed lavish potlatches that can last anywhere from a few days to several weeks.[41]

During a potlatch, the host will entertain up to several hundred guests who have been invited both from his own and from other tribes. Sometimes the host has prepared for the occasion for several years, during which he has gathered, bartered, traded, crafted, and stored the necessary food and wealth. When the festivities begin at last, the host is ready: He will treat his guests to the best meals he can muster, day in, day out. At the very end, he will honor his guests with the gifts he has prepared—blankets, paddles, masks, tools, musical instruments, clothing, and many others.

Potlatch culture spans from the subtropical coastal regions in the south, all the way into the far Arctic regions of the continent. Incidentally, this is the precise geographical range of the salmon. Biologist Jim Livatowitch observes that this might not be a coincidence, finding striking similarities between the life cycle of salmon and potlatch culture.[42] Potlatch events redistribute wealth from the most affluent members of a community to others both within and outside one's own community.[43] Those who have relatively more to give than others are held in high esteem for sharing their abundance with everyone else. It can take a host several years to prepare for the event, but once it begins, his munificence is unreserved and absolute. The descriptions of potlach culture actually evoke the life cycle of salmon—like potlach hosts, salmon bodies nourish all who live inside the watersheds: bears, coyotes, ravens, flies, humans, spiders, other fish, Sitka spruce, alder, maple, cedar. Like potlatch hosts, who often spend years gathering provisions, it takes salmon years to prepare themselves. When they do return, their gift is absolute, as abundant and munificent as the potlatch hosts.

It is unlikely that the similarities between the salmon's life history and Klallam culture are coincidental. Humans not only live side by side with the salmon—they have coevolved with them. The salmon who are now native to the Olympic Peninsula are of relatively recent origin; like the humans, they did not arrive here until after the last glaciers receded northward some ten thousand years ago. The memory held by the bodies of the Olympic salmon, the intuition they pass on from one generation to the next, has grown out of a continuous, dialectic engagement between themselves and the human community. Salmon and humans have become indigenous to the peninsula not despite each other, but through one another.

Not All There

In 1787, thirty-two-year-old Captain Robert Gray, a veteran of the Revolutionary War, was commissioned by Boston merchants to make a trip along the coast of the Pacific Northwest. He was to trade fur, then sail to China, exchange the furs for goods, and make the long trip back home to Boston. Natives were to receive trinkets for their furs. Boston merchants were to receive riches for their investments. As for Gray, he was to receive what sailors like him received—the recognition of history.

On that journey, Captain Robert Gray became the first American-born European to circumnavigate the planet. He returned from his expedition, and soon after embarked on another journey to the Pacific Northwest. This time he sought to open new fields for the fur trade and to further explore the still unmapped coast. In the east, merchants and statesmen alike were pushing an aggressive agenda of expansionism toward the west, and explorers were sent out to identify the best trade routes.

Like Juan de Fuca, Gray was unable to find a sea passage straight through the continent. But he did stumble upon a very large river, one of the Pacific Coast's largest. On the day he sailed into the river's mouth, he ignored the presence of the indigenous community who came to greet him. He never asked the locals for the name of their river. Instead, he called it the Columbia River, not after himself but after the ship he was sailing. To this day, popular accounts credit Gray as having "*discovered* the Columbia River which forms most of the northern border between Oregon and Washington State to the north,"[44] never mind the fact that, according to the same source, "there were over 100 Native American tribes located in the area."

Captain Gray kept a meticulous log of his southward journey along the Pacific Coast. He sailed past what is now known as Vancouver Island (a name commemorating another European sailor, the British Captain George Vancouver), past the Juan de Fuca Strait, and down the coast of the Olympic Peninsula. One day, Gray anchored in a bay that was home to the Quinault people. Cedar dugout canoes came out to his ship. The indigenous cautiously inspected their visitors. Gray invited the natives to come on board. They interacted with the crew. The whites gave them beads and hammers and iron tools and some blankets, and they returned the favor with some fresh fish and a bucket of clams. At the end, the parties bid farewell, and the natives left the ship again. From the perspective of the European-Americans, this was an inconspicuous event, one among scores of similar such encounters between explorers and natives. It did not take up much space in Gray's log. Gray and his men sailed on; the still nameless Columbia River lay in wait. European-American history bypassed this unremarkable meeting on the wind-beaten, rain-soaked edge of the continent. Nothing extraordinary had happened. This was just another chance encounter with the savages.[45]

But there is always a possibility that a stray memory might breach the surface of appearances, like a salmon in whitewater rapids. An observation tactfully left out, a story skillfully avoided, can even obliterate other people from living memory. And yet sometimes a memory will hold on in dormancy, in the shadows of the dominant narrative, refusing to go away. And from there it might breach, casting new light on the assumed order and relevance of events.

The ethnographer, teacher, and storyteller Chuck Larsen once decided he would take the logs of Captain Robert Gray to the present elders of the Quinault tribe.[46] Of both Norwegian and Native American ancestry, Larsen long nurtured a curiosity for the way stories of white expansionism are told from the other side—from the side of the disprivileged speakers. He read to the elders from Gray's logs. When he was done reading, Larsen asked them, "What is *your* side of the story? How have these events been handed to you through your oral tradition, for more than two hundred years?"

It became very quiet in the room. The elders looked at one another. One began to chuckle. Then another. Somebody said, *hoquat.* Then they all laughed wholeheartedly. *Hoquat. Hoquat.*

When the elders had caught their breath, they told Larsen the story the way it had been handed down to them, from father to father to father to father.

The Quinault realized that there was no physical threat on board the visitors' ship. It was peaceful on deck. The Quinault wanted to trade, and they wanted to have a good look at the ship. They stayed aboard for several hours. There was much to see, inspect, observe. When they were ready to go, the Quinault said farewell to their visitors. Then they left and rowed back to shore. When they landed, they said: What was that all about? These strange new humans are ugly! They smell! There was also a detail of their encounter that stuck out, a detail they marveled at: *Something was missing from these people.* They were incomplete. They were partial people. When you interacted with them, it was like seeing someone off in the distance, without any clear contours. You saw them, and you could not see them. Not really. Their details were obscured. It was as if these new humans were clothed in a constant veil of fog.

At long last, they found an expression that captured these observations, as well as the puzzlement they felt: *Hoquat*, they said. *Hoquat*. In their tongue it means *The people who aren't all there.*

The Quinault, members of a thick and heterogeneous web of indigenous societies that coexisted along the continental coast, had encountered a group of people unlike any they had ever traded with. They were left speechless. Finally, they succeeded in capturing the essence of the encounter in poetic reference. For this is what the word *hoquat* is—poetry. The word is nonjudgmental. It does not point a finger. It merely distills an experience to its essence: Something was missing from these people. These people weren't all there. For every subsequent generation, it was enough to simply evoke the one word, and all who knew the story would be able to relive the puzzlement. *Hoquat*.

Gray and his crew sailed on southward to discover and name the Columbia River. In Gray's perception, he and his crew were entitled to claim the rivers, the forests, the sea, the mountains, and all creatures as theirs. The act of naming things became their first act of ownership. In their perception, the existence of the indigenous was irrelevant. What mattered was their own narrative, which compelled them to pursue full-fledged imperial conquest, mapping the land according to the new historic imagination.

And so the Grays, de Fucas, Vancouvers, and Lewis and Clarks of the sixteenth, seventeenth, eighteenth, and nineteenth centuries sailed up and down the coast, naming things, mapping the terra incognita, bargaining, laying out plans for settlement, for development, and later, for industry. As they did, they acted out of the conviction that theirs was the only true way of seeing. Theirs was the only way of telling the story of the West. With their chosen names planted firmly in the newfound land, they were well under way to writing their own history in its empty pages.

But not all living memory will be deluged by the dominant narrative. Some bits of memory linger on in bright daylight, there for anyone to see. Half a century after the Quinault's encounter with Gray, the whites enforced treaties with the natives. The white settlement of Aberdeen in the southern part of the Olympic Peninsula moved across the river, and as it did, the native community was forced onto reservations. The natives began referring to the place they had been forced to give up as *Hoquiam—the place where the people who aren't all there live*. To this day, the Olympic town of Hoquiam retains that name.

When the River Spoke

Thomas Aldwell lay by the Elwha River bank and looked up into the sky. At that moment late in the nineteenth century, history's arrow was engaged in a fierce struggle with the voices of the land. Both were toiling to make themselves heard over the other. It almost seemed as if the land's voices would prevail. Just then, the river spoke to Aldwell: "My life had taken me to schools, to cities, to business, but suddenly that spring [at the Elwha] embodied all of life and beauty I thought I'd ever want." The experience of being drawn into the presence of so eloquent and beautiful a host of voices was so powerful that even fifty years later, when he sat down to pen his autobiography, Aldwell could still step into their immediacy. His recollection is saturated with river life, and with the life of the senses: "Those days packing into the Elwha cemented me to the Northwest. The needle-covered Earth, the patterned tree branches, the sky, the fresh bracing woods' smell—all seemed to make me part of the Earth I rested on." Aldwell had come from so far. When he actually heard the river speak to him, he had nearly arrived.

But his lifelong habit of living inside the notion of linear history drew him away from the land he so yearned to be a part of. He got pulled along by the narrative inertia of his lineage: "The property on the Elwha River always fascinated me, but it was not until I saw it as a source for electric power for Port Angeles and the whole Olympic Peninsula that it magnetized all my energies." The place had spoken, but his narrative screen filtered out the voices. A small window of opportunity had opened, and then it slammed shut for good. The dams became his life's calling. Never again did this industrious settler step into the full immediacy of the Elwha's presence. Never again was he fully there. Later he would say: "I sometimes wonder if I ever saw Port Angeles *present* because so much of my time was spent seeing Port Angeles *future* and as that future became present, I kept looking ahead."[47]

The dams were hailed as a milestone as progress swept into the land. When construction on the Elwha was about to commence in 1910, the *Port Townsend Morning Leader* published the headline "Cause for Congratulation" in its August 3 edition. Beneath the headline were these statements: "Any man in Port Townsend who knowingly throws as much as a straw in the way of the complete success of the Olympic Power and Development Company should receive the unqualified censure of a united community and should be consigned to the catalogue of undesirables of which this city in company with every community in the state of Washington already has too many. Here's to Tommy Aldwell and to his Olympic Power and Development Company! May they live long and prosper!"[48]

A year later, on May 1, 1911, the *Seattle Times* over on the mainland wrote proudly: "Far down in the canyon cut through the rock by the Elwha in its centuries-long course to the seas appeared the evidence of man's ingenuity in harnessing nature. . . . Up rearing high in the air, a gigantic derrick with mathematical precision lowered slings of Earth to the men at work on the coffer dam more than 100 feet below, while the roar of machinery vied that of the river. The . . . Elwha is now under control."[49]

For the Klallam, there was no reason to join in the festivities. The place of coiled baskets, most sacred of Klallam sites, their Mount Sinai, their Mecca, was not far upriver from the canyon where Thomas Aldwell had envisioned the Elwha converted into an organic machine. When the Elwha Dam went in, the water began to rise, inch by dammed inch of reservoir backwater. At last, the creation site was lost. The place of coiled baskets drowned beneath the influx of peace, power, and civilization.

A tension stirs in these pages. We see a certain aloofness of the historical imagination as the linear story moves across the articulate land and its articulate inhabitants. Its arrival in the Pacific Northwest leaves many creatures silenced, peoples marginalized, voices oppressed. The new humans seem rather alien to this place's dialects and poorly trained in listening to the land. The settlers are heirs of a narrative tradition that thinks the world ought to be remade through reason alone. These heirs wield a certain mythological power: Their belief in narrative exceptionalism gives them a sense of entitlement over the Earth and over those dwelling within it, including other humans. They begin to create technology, the likes of which had never before been unleashed on the Pacific Northwest, such as dams designed to be permanently in a river. Their narrative exceptionalism and their technology enter into a cycle of mutual reinforcement, each reassuring them that the land and its inhabitants indeed are silent and inert. They are storytelling animals, yet they seem to live more in their story than in the storied land. The narrative within which they have made their home is not entirely of this Earth. It is not quite, never fully there. It is a utopian story (a story "out of place"). It is abstract and idealized, easily uprooted, easily shipped to foreign shores. Yet it cannot really take root anywhere. It is the story of disembodied minds in permanent exile, attempting to dominate all that is perceived as "other."

But the land, under this sustained onslaught, ever struggles to be heard. Every boulder on the beach, every crested wave roaring toward the shore, every vulture soaring on a mountain updraft, every constellation of stars and planets in the night sky above spontaneously honors the speech of things. It is a speech from which no human community is exempt. It encourages the storytelling animal to pay close attention: Learn to hear! Strive to make sense of what you hear! Respond! Or else, face the consequences of your narrative amnesia.

CHAPTER TEN

Salmon Boy

*This living flowing land / Is all there is, forever /
We are it / It sings through us.*

GARY SNYDER

*I*t starts in autumn, when the leaves are falling. Humans might or
might not notice the imminent change of seasons when the trees
begin to shed their leaves. But the falling leaves do not go unheard. They
pound the ground with the rhythmic calling of drums, and the Earth
begins to vibrate with the pulse of the dying season. The pulse rolls out
to sea, where the salmon are. They have spent several years out in the
Pacific, traveling in a great circle, but when the drums start calling, they
swim home, eager to return and dance.

One day, a boy was playing not far from the river. He knew that down
in their underwater villages, the salmon shed their salmon skin and
turned into humans, and that when they did, they not only looked like
humans—they became humans. He knew also that the salmon could do
something humans could not do for themselves—the salmon fed them.
Salmon fed humans, and in giving themselves as flesh, they became
human. The boy knew—as everyone did—that this balance could be
easily upset. He knew that what the salmon did for his people was part
of an original agreement, and that if either side were to violate the terms
of it, the agreement would be broken, the cycle interrupted.

But that day, the boy decided he did not want to throw the bones
back into the river. He felt like playing somewhere else, and he did not
wish to walk all the way down to the water. So he dumped the bones
into a bush on the side of the little path. After that he went to play, and

when he came back he told his people he had returned the bones as he was supposed to. Everyone believed him, for no one had any reason not to. Everything seemed fine. But the next time the boy was playing by the river, he fell into the water. He started drowning and was sure to die, but the salmon came to his rescue. They were on their way back to their underwater village when they saw the boy in the water, struggling to make his way to the surface. They pulled him down into their village and fed him well. They kept him safe and treated him with all the privileges and comforts of a specially honored guest. He began to recover, and he played with the salmon children. He learned to love them all, young and old—they were people, just like him.

One day, as they were playing, the boy saw a girl. Half her body would not move, and he wanted to know why. He asked some of the other children about her, and they said, "She has no bones on that side of her body; her face, her arm, her leg, her ribs—there aren't any bones there." The boy said nothing, but he realized that she was suffering because he had not returned the bones to the water when he had a chance. These were the same salmon who had been in the river when he was playing there. They were connected directly to his community. They were a village of his people.

Because the salmon run only once a year, the boy could not return to his human relatives until a year had passed. When he went home, he told his people what he had seen. "These people took care of me," he said. "They honored me, and they respected me. And my actions caused sickness in their community. In many ways, they are more important than we are, they have more spiritual significance than we do, they have more power than us. And so we have to do this—we have to maintain our relationship. They respect us so much, and we have to respect them the same. If they honor us, that's what we have to do to them."[1]

———

The story of Salmon Boy opens the listener's experience to being embedded within a living Earth teeming with agency. It describes a moment of transition between two generations. This is a moment of peril: The string of indefinite dwelling in place can be severed. The knowledge can get lost. The boy does not grasp the long-term implications of his refusal to return the bones. His imaginative horizon is too narrow for him to fully understand the original agreement. He lacks a certain motivation to close the cycle of mutual obligations. And in his ignorance, he causes suffering. The peril is never abstract; it is always close at hand: The cycle of knowledge can unravel. If his generation does not learn to fully embody the original agreement, then the web of participation will loosen, become fragmented, and disintegrate.

The boy started drowning. This is a psychological transition: His drowning is not only a drowning of the body but also a drowning of the spirit. In his ignorance, he is lost, disoriented. His obliviousness threatens to kill him.

But the salmon came to his rescue. Salmon Boy is not penalized for his misdeeds. When he enters the house of salmon, he is not treated as an offender but is received with the highest honors. The salmon speak through their actions: The boy must see with his own eyes the suffering he has caused. To sanction him without offering him the possibility to experience the direct consequences of his actions would not guarantee his understanding. He is neither punished nor prematurely forgiven: He is given an opportunity to redeem himself. Only then will he be able to ripen to maturity. Budding into maturity, he will rediscover the crucial distinction between the power over others and the power to be in relation.

The psychologist Mary-Jayne Rust has described archetypal stories of psychological passage in which a "young male hero who battles against nature" undertakes this task by "trying to rise above everyone and everything, but in doing so he loses touch with his earthly nature, his bodily instincts and intuition." Though the story of Salmon Boy is not exactly of the same type, the required lesson is similar: The hero's "ego needs the teachings of intimate immensity." He must become a hero "who is able to die and be reborn, who knows of 'power with' rather than 'power over.'"[2] The story of Salmon Boy can be read as the death of the ego—symbolized by the boy's near-drowning—and the subsequent rebirth into a more comprehensive sense of self. Here, the salmon step into the picture. This opportunity for the boy to succeed in his crucial rite of passage is a gift given to him by the salmon. It is by their choice—it is thanks to their agency—that he emerges into maturity. They assume the role of elders without whose guidance his psychological development would remain stunted.

The recognition of the salmon's agency also has other implications. It is inconceivable for the Coast Salish to think of salmon as "resources," or as "stock," or even to think of them in terms of the subject-verb-object convention of "we-catch-fish." This has direct implications for the way Klallam economy is organized. Unlike in the modern tradition, salmon are not treated as passive objects; humans do not merely "catch" them (or, more recently, "produce" them). The salmon *gift* themselves freely and willingly. They *choose* to give themselves as food. And when humans have consumed the flesh of the fish, they are expected to gift back to the salmon their bones. Lewis Hyde observes that it is this returning of the bones, so common throughout the Pacific Northwest, that "makes this truly a gift cycle."[3] But it is a cycle that comes with an obligation. Only when the recipients of the gift keep it circulating will it ultimately come round again in the future. "[T]he myth declares that the objects of the ritual will remain plentiful *because* they are treated as gifts," Hyde writes. "Gift exchange brings with it, therefore, a built-in check upon the destruction of its objects; with it we will not destroy nature's renewable wealth except where we also consciously destroy ourselves."[4]

Note also the shapeshifting quality of these salmon. The story tells us that they can become human! A human child lives an entire year together with them. The salmon live in underwater villages, which are just other villages of the human community. When the boy and the salmon speak with one another, they do so fluently and effortlessly, without getting lost in translation across species barriers.

Vine Deloria has written about the importance of shapeshifting in tribal religions. "Very important in some of the tribal religions is the idea that humans can change into animals and birds and that other species can change into human beings," Deloria writes. "In this way species can communicate and learn from one another." Though this shapeshifting has often been misunderstood and classified as "witchcraft" or sorcery, Deloria notes that "what Westerners miss is the rather logical implication of the unity of life."[5] Deloria points out that in such tribal religions, including the Native American tradition, there is no categorical difference between the different creatures.

What makes sense in indigenous thought also makes sense in phenomenological thought. Phenomenology describes knowledge as a fluid function of inner experience and outer events. The "world" of phenomenology is a participatory field of experiences co-constituted through a near-infinite array of different angles, actors, co-creators. As we have seen, Abram's writing has made this shapeshifting quality more accessible to those brought up into the Western canon. "Incomplete on its own," he writes, "the body is precisely our capacity for metamorphosis. Each being that we perceive enacts subtle integration within us, even as it alters our prior organization. The sensing body is like an open circuit that completes itself only in things, in others, in the surrounding Earth."[6]

Salmon Boy's shapeshifting is not so very far-fetched. If human speech is fully continuous with the eloquence of the air, water, and mountains, if our voices are but a certain style of Earth speech, and if our human thought is fed by the nourishment of wind, water, soil, salmon—if, that is, we not only speak of the land, but if the land speaks (or sings) itself through us—then speaking is not an act that holds us apart. Rather, it allows us to participate more virtuosically with the land's immanent articulateness. And the land also sings through all other creatures, as well as through river, mountain, and rock. All voices are Earth voices—regardless of whether they possess central nervous systems like ours, or whether they have the decentralized sentience of trees and fungi, or whether they are fully inorganic beings such as rock, water vapor, or hydrogen atom. The shapeshifting quality of the story of Salmon Boy is a nod toward our ability as humans to become mindful participants in the speech of things, to keep our improvisational intelligence open to other intelligences and articulate styles, to hold our perceptions semipermeable and receptive to other sensuous bodies, each of whom is also participant with the larger Earth body.

Like Children

In 1789, Friedrich Schiller gave the inaugural lecture for his new professorship of history at the University of Jena. In the lecture, he spoke of the indigenous peoples with whom European travelers were coming into contact during their journeys. To Schiller, all of them were like children: "The findings which our European sailors . . . show us societies arrayed around us at the most varied levels of development, like children of different ages might stand around an adult, and through their example they remind him what he himself formerly was, and from whence he departed."[7] A century and a half earlier, Francis Bacon struck a similar tone when describing the indigenous peoples of North America, which he referred to as New India: "Let a man only consider what a difference there is between the life of men in the most civilized province of Europe, and in the wildest and most barbarous districts of New India; he will feel it be great enough to justify the saying that 'man is god to man.'"[8]

In the modern narrative, indigeneity has been commonly imagined as an "earlier" or "lower" state of humanity, one that inevitably will be overcome. In the 1908 *Encyclopedia of Religion and Ethics*, the entry on "Animals" includes the following:

> Civilization, or perhaps rather education, has brought with it a sense of the great gulf that exists between man and the lower animals . . . In the lower stages of culture, whether they be found in races which are, as a whole, below the European level, or in the uncultured portion of civilized communities, the distinction between men and animals is not adequately, if at all, recognized. . . . The savage . . . attributes to the animal a vastly more complex set of thoughts and feelings, and a much greater range of knowledge and power, than it actually possesses. . . . It is therefore small wonder that his attitude towards the animal creation is one of reverence rather than superior.[9]

The notions of "lower animals" and of "lower stages of culture" have linear historical connotations. They both denote an Other of a supposed lesser sophistication than oneself, someone who was left behind, an inferior. The perceived gulf intrinsic to humanity-as-separation also implies that, at some point, modern humans successfully elevated themselves not only above other-than-human expressions of life, but also above other-than-modern expressions of humanity.

In the Salish stories, which the Klallam share with many loosely related coastal nations, "those who act like children" are not an Other but are members of their own community. They are not an earlier "stage of development" that has been overcome once and for all. They are a problem that keeps coming round and round! The problem of immaturity, for the Coast Salish people, is not a historical problem, nor is it an epistemological one. It is a psychological problem. And the most vital issue that demands their mature response is the very problem of indigeneity.

Indigeneity is the difficult and never fully finished skill of living in place. The problem of living in place cannot be solved by holding reason in exile and by assigning to reason the sole responsibility of composing an ever-more-perfect understanding of how to inhabit that place. It can neither be framed as an epistemological nor as a historical problem. It is an issue of psychological immaturity, either of individuals or communities; it is not a question of "less advanced," and therefore "immature" cultures.

This is because, firstly, indigeneity is an adaptive skill. It needs to remain receptive to the fluid vitality of places. Learning to live in place involves gleaning thought-structure from the patterned animation that is specific to each place. The metaphors for a community's relative success or failure cannot be imposed haphazardly onto the land, which is what concepts such as "modern" or "developed" do. Instead, the metaphors need to be modeled in direct, reciprocal participation with the land. Rather than being placed onto the land, they must be composed through a dialectic mediation within it. Only when the land thrives can the people who dwell within it thrive. If the people become a disturbance, the ripple effects will sooner or later come rolling back over them and seize their communities as well. The pathologies will enter into cycles of positive feedback. Reason, if held in isolation from the body, can hardly entertain a lasting, dialectic exchange with the living land.

Indigeneity also requires a certain continuity over time. The adaptive skill to dwell in place decently and indefinitely is always just one generation away from being lost. Continuity cannot be taken for granted. Each generation of elders must carefully attend to the younger generations as they gradually assume more responsibility for their own actions and for the community at large. This is foremost a psychological task: The young must be instructed so that they will develop a comprehensive sense of community, an identity that integrates not only their most intimate family members, their society, and humans, but also the very rocks, mountains, wind, and ocean. If the young ones are not helped to extend their sense of community deeply into the sensuous terrain, then that young generation will suffer a stunted fruition. It will not ripen into the fullness of its potential. It will remain incomplete, fragmented.

This challenges that particularly modern epistemological optimism. Its programmatic movement toward the ever-new directly renounces processes involving continuous, cyclical renewal. Modernity unbends the attention from cyclical processes into a line. Linear concepts like "progress" or "development" make it difficult to conceptualize the cyclical problem of individual and collective maturity. The cultural knowledge and social structures needed to live deeply inside the land, as one remark-able creature among many, are not won once through a succession of reason-derived deductions and then possessed ever after. Indigeneity, as a collective effort, requires periodic renewal, rejuvenation, recomposing. It is hard and risky work.

But when that work succeeds, it is possible for humans to live in place in a state of indefinite, dynamic maturity. Humans can become mindful participants in the land's larger unfolding. They can dwell creatively without becoming a disturbance. So here are at last some signs of epistemological sophistication: To weave such an understanding expansively into one's cultural web. To code the necessity of living in place and the psychological difficulty of maintaining that awareness into functional social practice. To apprentice with elders both human and other-than-human. To allow Earth voices to slip into the midst of the human community. To let them linger so that they might instruct humans when they are in need of a more comprehensive perspective. And finally: To take rightful pride in the power of reason but stay mindful of the need to bring it back into participation through the life of the body.

Becoming Native to These Lands

Paleoanthropologists observe that humans first migrated into the double-continent of the Americas after the last glaciers had retreated back to the north. Never before had the other creatures seen human footprints on these beaches or heard their songs weave into the canopy of these ancient forests. The native peoples of the Americas had not always been native to these lands. They, too, had once been new humans, aliens who hardly knew the adaptive skills it would take to thrive here. They did not always know how to live here without diminishing the inhabitants who had been here long before them.

It is a well-established fact that the arrival of humans to the Americas coincided with an extinction wave of much of the double-continent's megafauna. Precisely what role humans have played in this extinction wave continues to be a matter of debate, but their direct impact is at least "plausible."[10] The geoscientist Paul Schultz Martin argues that "virtually all extinctions of wild animals in the last 50,000 years are anthropogenic, that is, caused by humans."[11] Schultz Martin draws a vivid image of North America's fauna before the arrival of humans, suggesting that the continent was once home to a far greater variety of now forgotten big animals: "North America lost mastodons, gomphotheres, and four species of mammoths; ground sloths, a glyptodont, and giant armadillos; giant beavers and giant peccaries; stag moose and dwarf antelopes; brush oxen and woodland musk oxen; native camels and horses; short-faced bears, dire wolves, saber-toothed and dirk-toothed cats, and an American subspecies of the king of beasts, the lion. . . . Before extinction of our native big mammals, the New World had much more in common with an African game park."[12]

If indigeneity can be understood as a creative adaptation, a fitting-in-place through the prolonged habit of attentiveness, then the loss of America's megafauna could be partly due to the fact that the newly arrived had not yet sufficiently crafted cultures appropriate for these lands. There was creative work to do, and success was by no means guaranteed. As was true on other continents into which these two-leggeds wandered, loss was a stubborn companion to that work.

———

Indigeneity is not a status you gain once and then possess ever after, nor is it a title one holds. It is a competence, a capability, expressed through social structure. It is the process of fine-tuning your presence in an ongoing dialectic with the places you inhabit. You discover them, and then you discover them again. And then you negotiate and renegotiate the terms of your presence with those who dwell there alongside you—with salmon, otter, whale, raven, with other human communities.

Indigeneity emerges out of trying and failing, again and again. You will fail and fail, and then you will discover that through it all you have found a way to coinhabit the place with everyone else. You have found your place. You and those around you have come to a mutual agreement you can all live with, quite literally. At that point you might discover that your presence among those who are unlike you—among those who speak a different human dialect, those who sprout leaves in the spring, those who hibernate beneath a protective blanket of snow, those who dig dens into the gravel bank of your river, those who hold on to these banks with the numberless fingers and fingerlings of their roots—is no longer a disturbance. It is the way things are. Like everyone else, you, too, are fully there.

Your group of kin have inhabited a place so observantly that you have gradually become fluent in the land's speech. A community of native speakers emerges through what restoration activist and poet Freeman House has eloquently called "the long practice of cumulative attentiveness."[13] This is a community born into the articulate and richly stratified depth of the land—a community birthing itself into that depth. An indigenous community will have been in place long enough that it has begun to see the innate tendency of all things for change. It will bear witness as these tendencies arrange themselves in different patterns of flow. And it will strive to carefully translate these patterns of flow into culture: to let the innumerable rhythms and patterns of the sensuous terrain resonate more fully in human thought, and to let the land's speech emerge more audibly into human language.

Indigeneity involves the open-ended cultural work of striving to integrate the storytelling animal into the shifting depth of the living terrain.

Rather than a remnant of the past, rather than a certain "developmental stage" in our ongoing creative adaptation as human animals, and rather than an ethnic category, indigeneity describes a lived quality that is possible anywhere, at any time. More than that, it describes a quality of participation with Earth that is necessary for any community, if they wish to endure within the storied unfolding of a fully animate, living planet.

Breaking Through the Cycle of Abuse

The people of the Elwha were repeatedly drawn back into the thick of the land by their oldest ancestors, the salmon. Klallam culture was incomprehensible without a certain felt reciprocity between humans and salmon, between humans and the larger river community.

Klallam identity had its roots in the water, the soil, the air. In the Klallam universe, personal, cultural, and natural phenomena formed an integrated whole in which each dimension constituted and gave meaning to the others. To the Klallam, maturity was expressed by a person's ability to affirm her place within the kinship bonds of killer whale, western red cedar, elk, black bear, cougar, and, of course, salmon. Each of these was spoken of as other people, no matter whether feather-clad, slick-skinned, or fur-coated, whether they sang their songs out in the ocean, wafted alertly on warm updrafts with wings spread, drank from the soil with rooted limbs, or hurled themselves up whitewater rapids.

Affirming one's place in the community could come at great personal risk. To receive the powers of the ocean, young Klallam would sometimes paddle out with their canoe, lift a big rock, and jump into the ocean where they knew it was the deepest. The rock would take them to the bottom. Not all who let themselves be pulled to the bottom of the ocean would live. But those that did were taken to the houses of the ocean dwellers, including the houses of the salmon. There, they would be given their power. They would be shown what they were supposed to be able to do in life. Andrew Fitzgerald shared this ritual with me. He also said: "Your life, when you grow up, means nothing until you have your power . . . people are impoverished without it . . . like with the idea of living in poverty . . . the only way that our people ever considered something pitiful or impoverished was when they didn't have a power. . . . *Power* is a hard word to translate into English. What that means is that you didn't possess anything that would be able to . . . that comes from within your community around . . . that would be able to translate and help your community."[14]

This was the first time I encountered the possibility of a power relationship other than the one I was familiar with—the power over others. It was here I glimpsed that to be truly powerful might mean having power in relation to others. Power need not be a hierarchical concept; it can be a relational one. Everyone in the Klallam community had a different power, each bestowed on them by the land that nourished them. Some were warriors. Some were peacekeepers. Some were healers. Some were diplomats. Some were leaders. Some were storytellers. Everyone could do something that differed from what anyone else could do. In that sense, everyone had to be able to interact with their kin, for everyone held a piece of what constituted the community in its entirety. The powers of the ocean were held in greater esteem than most others, because those who possessed such powers had risked their lives for them. They had risked their lives for the community, searching for the power that resonated most strongly with their own identity. They had returned alive, and they had been reinforced, fortified. They could now lend their new-gained power to the benefit of all.

During our conversation I asked Andrew Fitzgerald whether he would translate the Klallam word for *human* into English for me. He thought about my question for a while, groping for words that would help me understand. He said it is not an easy question to answer. Many words have come to his people after contact, and these new words have distorted the earlier observations to a degree. "But basically," he said at last, "the way we describe it, is like, *we're the most pitiful*, right, of all the beings . . . It's a long process. . . . But it describes us. It's more of a descriptive word. And it means, '*by the pitifulness of our thinking.*' . . . We can't fly like the birds. All we can do is sit here and think about things, and it causes us to be able to cause problems within ourselves."

I told him that the philosophical tradition of the moderns maintains an observation that is simultaneously similar and very different. I mentioned Descartes, who was born in the same decade that Juan de Fuca explored and named what he perceived to be deserted lands. I mentioned that for the past four hundred years, our philosophical tradition has been deeply informed by Descartes's *I think, therefore I am*. Being human, for Descartes, meant two things: First, it meant possessing a faculty no other creature possessed—thinking, the rational mind. Second, it meant that this faculty not only separated, but elevated humans above all other animals and above the rest of the vast, complex, and rich community of life.

So there we were, he and I: progeny of two cultures who had observed, independently, the remarkable power of human thinking. Each of our cultures had forged this discovery into language and then carried it on through philosophical tradition. And though both of our cultures had long ago made the same observation, the ways each interpreted the observation were astonishingly asymmetrical. In my people's interpretation, it went: *I think, therefore I am separate and superior*. In his people's interpretation, so remarkably similar and yet so different, it went: *I think, therefore I am pitiful*. In our tradition, it might have sounded like this: *Cogito,* ergo *miser* sum.

The Klallam couple their observation of humans' power to think with the humbling insight that this power, like others left unchecked, can become a source of corruption. They are alien to the modern insistence that rationality would, by itself, justify an absolute difference between humans and other creatures. Experience has taught the Klallam that humans are one among a multitude of uniquely gifted people. Some creatures are faster than they are, some are better able to stalk prey, some have keener eyes, and some have more sensitive ears. There are bird people gifted with the ability of flight. There are fish people gifted with the ability to inhabit the deep seas. There are tree people gifted with longevity. Experience has taught them to acknowledge every one of these creatures for what makes them unique. In a world as diverse as the Elwha River basin, the most commonsensical and plausible interpretation of their own power—the power to think—is not that it elevates them above that world. It is, rather, that it demands of them extra care to not become too alienated.

Their recognition of the astounding gift that is human thought also resonates with a keen understanding for the gift's shadow aspect. Thought endows humans with certain freedoms of action that are not given equally to other creatures in the land, and while these freedoms of action are a thing of beauty, there is an ugliness that looms ever near. The radiant light of thought can illuminate. But thought can also cause suffering. The possibility of corruption ever lurks in the shadows.

There is no comparable integration of this dark side of thought in the ratiocentric tradition. Given that the modern narrative is marked by such unchecked epistemological optimism about the power of thought, Klallam philosophy is more than just its opposite. Klallam philosophy might be *more reasonable* than its reason-centered counterpart. The story about the genesis of the Salish universe suggests that self-aggrandizement at the expense of others is a manifest sign of immaturity. The first people are like children. They never grow up. They begin comparing themselves to others. Driven by the craving to distinguish themselves from everyone else, they become narcissists. They begin living at the expense of everyone else. This is a rather astute description of the modern condition and of the suffering endemic to it. It prefigures the work of Paul Shepard, whose *Nature and Madness* drew links between the systematic destruction of life and a culturewide, stunted maturation of the collective. In Shepard's thinking, moderns are denying themselves the possibility to fully mature.

According to the Salish analysis, the decisive impetus for breaking through the vicious cycle, where suffering leads only to more suffering, comes from the salmon: It is when they at last consent to return that reconciliation becomes possible. Their return is a promise that comes with an obligation: Only when all subscribe to the original agreement will the suffering end. The original agreement marks the moment when the parties step across a pivotal psychological threshold: In signing up to honor the salmon in perpetuity, the children step out of their narrow adolescent horizon and into a far larger, and far more integrated, sense of identity. They emerge fully into the larger community to which they are beholden. Their sense of self is now inseparable from the living land. This is the moment when maturity is reached. It is this process of reaching maturity that helps break through the psychopathological cycle of violence.

Technologies of Inhabitation

There is widespread agreement among historians and anthropologists that Native Americans had both the knowledge and the power to overshoot their economic base.[15] The weirs, nets, and weapons they crafted could, in principle, have killed salmon runs in a few seasons. The Klallam knew how to close off entire rivers to catch spawning salmon. But evidence suggests that the interwoven web of salmon-based Native American people throughout the Pacific Northwest, including the Klallam, never seriously overshot their economic base, nor did they imperil their own livelihood. In fact, at the time of contact, numbers of spawning salmon throughout the region were robust and, as far as the record goes, at their historical peak.[16]

One Ezra Meeker recalled the sight of salmon at the headwaters of the Puyallup River: "I have seen the salmon so numerous at the shoalwater of the channel as to literally touch each other. It was utterly impossible to wade across without touching the fish."[17] When Lewis and Clark arrived at the Pacific after their trek across the continent, they found such a "multitude of the fish" as to be "almost inconceivable. The water is so clear that they can readily be seen at a depth of 115 or 120 feet. But at this season they float in such quantities down the stream the Indians have only to collect, split and dry them on the scaffolds."[18]

The anthropologists Campbell and Butler have studied the region's catch record from the past 7,500 years.[19] They conclude that throughout this vast stretch of time, humans lived sustainably in these lands, despite changes in ecological and social conditions. The climate changed gradually, sea levels rose and changed the lower stretches of rivers, and seismic activity and volcanic eruptions punctuated the felt continuity of time repeatedly. Meanwhile, the population seems to have grown, leading to increased pressure on the land's many gifts, salmon and other.[20] Yet the land went on teeming with fecundity. "Overall, the proportion of salmon taken relative to other fish changed little over 7,500 years," write Campbell and Butler, continuing that "this indicates sustainable use, rather than overfishing, despite the fact that the region supported extremely high human population densities, and cultures had the technology to greatly reduce salmon populations."[21]

Here we see glimpses of the emerging skill to dwell decently and indefinitely in place, of the continuous process of culture building that is not static but constantly fine-tuning itself, and of people not passively adapting to external conditions but actively fitting themselves into place, creating resilience both within the web of human interactions and throughout the web of more-than-human participation. The result is lived indigeneity. It seems possible to maintain large numbers of salmon even under extremely high population densities, and for the mind-boggling span of more than seven millennia.

This point is critical. Without it, it would be hopeless to study the epistemologies and social contracts of other indigenous people for clues about how to compose contemporary expressions of indigeneity, ways of being of this Earth that are meaningful given our contemporary epistemological horizon. The historian Jeff Crane writes that "the Indians had the technology and the motivation to harvest far beyond their own needs for consumption. So, what stopped them from doing so? . . . *a spiritual and taboo system emphasizing respect,* propitiation, fear, and balance limited salmon harvest and destruction."[22] The pervasiveness of healthy salmon suggests that there were intact social structures in place that not only fostered knowledge of technology's propensity for overshoot, but that effectively translated such knowledge into a functional ethics.

———

The technology the Klallam crafted, like their stories, resided entirely in the land. It emerged from the forests and waters and plains. When it no longer served a practical purpose, the Klallam let it return to the land. It became soil again, food for those who would come later. Far more than determinate objects, the tools the Klallam crafted translated their assumptions and beliefs into tangible action and served as expressions of their storied existence.

Martin Heidegger's work has shown that technology can never be thought of merely as tools. It is part of a social system that springs from and yields a particular relationship with the world. If technology embodies social systems, then the kinds of technology we use are indicative of what our society values, of what we deem true and what we deem false. The same is true of the Klallam. The technology they used was indicative of their most deeply held values. The Klallam were a powerful people with powerful technology. They knew how to fish birds out of midair with a bird net the height of a modern-day ten-story building. They knew how to intercept hundreds of thousands of salmon as they swam against the Elwha's mighty current. Using controlled fires, they knew how to re-create a whole landscape from a climax forest community into grassland. They had the knowledge and the skills to fashion an entire material culture—food, housing, musical instruments, garments, weapons, boats, artwork, ornaments, medicine—from within the thick of the living land in which they dwelled.

Yet they remained alert to the double-edged power of technology. Through narrative, rituals, and taboos, they retained the knowledge that every piece of technology gave them an advantage over those who depended solely on the faculties of their own bodies—such as bear, cougar, or salmon. They were also aware that with greater power comes a greater peril of corruption. Technology must be held in check lest its power turn against its own maker. At its best, technology would be integrated into the participatory relationships that all creatures of all nations engaged in, throughout the watershed. At its best, every aspect of Klallam culture—including technology—would nourish those who nourished the Klallam in return.

Klallam practices—their stories, taboos, rituals, repercussions for bad conduct—were part of a lacework of rites and regulations that spanned much of the North American continent, across kinship, tribal, and linguistic boundaries. There is a picture called *Snowshoe Dance* by George Catlin from 1835. It shows a group of ten dancing Ojibwa men, all adorned with festive headdresses and snowshoes. They dance in a circle around three tall rods that are spiked into the snow. All three rods are decorated with feathers, and the tallest of them displays an extra pair of snowshoes. The scene is dynamic and festive. The men sing and dance, and Catlin painted it so well that one can almost sense the pulse of rhythm in the frozen winter air. Catlin wrote about the ritual he had painted: "The *snow-shoe dance* . . . is exceedingly picturesque, being danced with the snow shoes under the feet, at the falling of the first snow in the beginning of winter, when they sing a song of thanksgiving to the Great Spirit for sending them a return of snow, when they can run on their snow shoes in their valued hunts, and easily take the game for their food."[23]

This greater ease came at a price. Their snowshoes made the Ojibwa faster runners and more successful hunters. It also distorted the odds between hunter and prey to such a degree that it brought about a serious risk: overkill. If they did not find cultural practices that would integrate their increased power meaningfully with the needs of the land—if they did not learn to adopt habits of effective restraint—that greater power would turn against them. Snowshoes gave them the means to kill as much as they wanted. But because of that, they also gave them the possibility to kill everything. Therein lies the deeper rationale for the snowshoe dance. It was not only a celebration; it was also a mindful act of reinstituting, year after year, the conditions of effective restraint: Act with forbearance! Show moderation! Observe the very real possibility of becoming corrupted! Note your capacity for being insatiable—and then reject it! The purpose of the dance was to subject technology, and by extension power, and by extension human ingenuity, to a patient regime of moral review and practical self-discipline. Without a voluntary and effective limitation, the blessing of the snowshoes had the potential to turn into a curse.

The botanist Stephen Harrod Buhner has observed that most indigenous humans, "living deeply embedded in their 'environment', all practice ceremonies and rituals that affirm and nourish the interconnectedness, the interbeing of the human tribe with the rest of the Earth family. This would indicate that the propensity to lose this connection is not just a modern phenomenon but is rooted deep within our humanity."[24] Likewise, it appears as if the celebration of interconnectedness is insufficient. It appears as if these practices are often coupled with the equally essential reminder to act with restraint and to prevent technology from running out of control. More often than not, the celebration of interconnectedness was also a check on technology's double-edged power.

The Ojibwa and the Klallam seem to have observed that technology—any technology—creates a leverage far greater than a person's bare hands. Increased leverage is what technology *does*. It innately magnifies a person's abilities. The long lines that the Klallam used, for example, extended the range of their bodies beyond the reaches of light and far into the deep sea. Nets as tall as ten-story buildings gave them an advantage even over creatures gifted with flight. The weirs, presumably the most powerful of all Klallam technologies, allowed the people to block entire rivers, and thereby to kill all salmon runs in the matter of a few seasons. Of course, they knew all this. And because they also knew to make the distinction between power over others and power in relation to others, they did not stop at the knowledge that the technology they created gave them power over others. They were not easily seduced by the kind of epistemological optimism intrinsic to the story of modernity. They took the next step and deliberated on the moral implications of their inflated abilities in relation to others. And thus, they understood that there is a connection between ability on the one hand, and responsibility—literally, "the ability to respond"—on the other.

In his book *Life Is a Miracle: An Essay Against a Modern Superstition*, Wendell Berry writes an analysis of modern science and technology that reads like a description of how the Klallam related to their tools. It is also a mortifying critique of modern technology. "What I am against," Berry writes, "is our slovenly willingness to allow machines and the idea of the machine to prescribe the terms and conditions of the lives of creatures, which we have allowed increasingly for the last two centuries, and are still allowing, at an incalculable cost to other creatures *and to ourselves*. If we state the problem that way, then we can see that the way to correct our error, and so deliver ourselves from our own destructiveness, is to quit using our technological capability as the reference point and standard of our economic life. We will instead have to measure our economy by the health of the ecosystems and human communities where we do our work."[25]

Berry's words not only capture the social meaning of the snowshoe dance. They also help us understand why the Klallam purposefully included holes into the bottom of their river weirs, allowing the salmon to pass through.[26] They help us understand why the Klallam never invented drawdown technologies in the first place—technologies that took more from the land than they were able to return, parasitic technologies that externalized the duty of paying for the cost accrued to other places, or to future generations. Instead, they utilized what the West Coast philosopher Derrick Jensen calls technologies of inhabitation.[27]

The Klallam understood that the question to ask of technology is this: How do we learn to use it in such a way that it does not diminish ourselves or the land we inhabit in the long run? How will we incorporate technology's innate tendency to amplify our leverage and our power over others into the necessity to immerse our lives into the creative processes of the places we inhabit? How do we integrate technology into an epistemology of uncertainty rather than let it be run by a boundless epistemological optimism? How do we allow for a generous precautionary margin between our actions and our inability to comprehend the full costs of our action?

All technology requires a dose of collective maturity. Moral responsibility follows from power, and power without responsibility begets exploitation. The original agreement of the Salish people says you may eat, be healthy, and enjoy abundance, so long as you honor your responsibility to us, the salmon. Nurture the relationships we have made possible, and do not forget the pledge you have made. Honor us with your lives. Defend us with your deaths. Kill us when you need to, but do not let your capabilities alone define your actions. When you do, we will all suffer. Create technologies of inhabitation. Tell stories of inhabitation. Learn the language of the river. Ask how you can thrive alongside us. Strive to emerge more fully into indigeneity.

Thanksgiving

The Norwegian archaeologist Lars Reinholt Aas documents the following stories:

1. "One of the classical Old Norse stories tells of how Thor was out riding with his goats. When night fell he sought rest at a poor peasant farmer's house. Thor slaughtered both of his he-goats, skinned them, and cooked them. He invited the

farmer's family for dinner—on one condition: They were not
to destroy the animal bones, but place them back into the
goat skin unscathed. Which they did, with the exception of
Tjalve, the farmer's son. He broke one of the goat's hambones
to get to the marrow. When the meal was over, Thor lifted
his hammer over the goatskin, and the goats rose whole, and
pulsing with life. But one of the goats was awfully lame in one
of his hind legs. Thor knew what had happened, and as an
atonement he took the peasant's son Tjalve and his daughter
Roskva with him. They both have followed him since then."[28]
2. "A traditional Dardic story tells of the encounter between a
hunter and the gods in the mountain. The gods invited the
hunter to their home. A mountain goat was slaughtered in the
hunter's honor, and he was assigned a piece of the hambone.
When they had finished eating, the hunter broke the hambone
to eat the animal's marrow. The gods became furious and cried:
"do not destroy the bones!" The hunter knew no other way
out than to replace the ruined bone with a branch. After the
meal, they all gathered the mountain goat's bones inside the
goatskin . . . , and the animal rose again, alive as before."[29]

The first story, of course, lives inside the land that is my home, Norway.
The other story is part of the collective memory of very distant relations
of the Old Norse people, the Dardic-speaking people who migrated into
relatively isolated valleys of the Himalayas, somewhere between 3,500 and
4,000 years ago. Aas suggests that it is difficult to think there is *not* a
connection between the two stories, a primordial myth shared by the common
early Indo-European ancestors both of the Dardic-speaking people
and of those distant cousins who settled on Europe's far northern rim
and became the Old Norse ancestors of today's Scandinavians. Given the
strong parallels, a common early myth or story seems very likely. Perhaps
the story of Salmon Boy, too, has its distant roots in that shared myth?

All of these stories seem to express a shared prototypical experience,
a pancultural knowledge. In each case, we encounter a polytheistic community
of humans, a people whose sense of the sacred is not gathered

around a singular metaphor but is dispersed widely and complexly throughout the land. Each is a community of people keenly aware of the multiplicity of personified voices in the land. Each is an animistic community of speakers living inside a land where all things are felt to be gifted with speech. In such a landscape, one is never truly alone. One is always being listened to, called upon, and summoned by forces larger than oneself. And one is held to account for one's deeds. One must tread with care through such a sensuous land: It is benevolent and bestows its gifts upon those in need, but it does have a dark side. It can indeed withdraw its benevolence. It can shapeshift from caretaker to nemesis. One must refrain from insulting the land's many animate powers, lest one invoke their fury.

There is also what we could call an ecological dimension to this animistic or participatory experience. It is in the way these more-than-human powers demand that people *give thanks* for the offerings they have received. It is this thanksgiving that keeps the original gift moving, making sure that the wheel of mutual participation spins on and on. Lewis Hyde writes: "Where we wish to preserve natural increase, therefore, gift exchange is the commerce of choice, for it is a commerce that harmonizes with, or participates in, the process of that increase."[30] In cybernetics, such cycles of participation are known as feedback loops. One agent *feeds back* to another an original impulse, a primary nourishment or gift, so that all who are in the loop enter into dynamic participation. It is a thanksgiving. And so long as the cycle of participation is not broken—so long as no one betrays the contract of mutual obligations—the cycle will not disintegrate.

This possibility of disintegration is ever present. Yet the cycle of the gift has some resilience. Earth voices all around us continue to summon us, beckon to us, subtly call out to us, entice us into noticing their presence. We are never fully alone or cut off. Even in our modern lifeworld we are persistently being "drawn back by the primacy of relationship," as Neil Evernden has expressed it.[31] There is *something*, or *someone*, drawing us back. It is the salmon, of course, and all the other critters. It is the rocks, and the clouds, and the rain. It is the gravitational pull of this larger body within whom we live. All continue to persistently lure the human animal back into the primacy of relationship.

. . . It Sings through Us

Indigeneity can be thought of as an increasingly competent and complex animistic participation between people and the land they inhabit. It is a collective competence, and a process with no shortcuts, an adaptive skill that requires a certain continuity over time. Some Earth voices withhold themselves to the casual observer. Others might withhold themselves even to a full human lifetime spent paying close attention, only making themselves heard as the added alertness of several generations begins to resonate together—through narrative, through speech, through collective habits.

Human communities become indigenous to their particular local Earth tongues through the practice of long and cumulative attentiveness. As a community becomes more sensitive and receptive to the land's patterned speech, further voices rise from the land, and the conversation gradually grows more involved, more reciprocal. Thought itself grows richer with birdsong, with river voices, with mountain endurance, with metaphors gleaned from migrating fish, changing weather patterns, and seasonal flux. Laws, institutions, traditions are shaped that honor and defend the land's complexity. Technologies are crafted that do not rip humans out of the finely woven loops of participation that compose the biosphere. "Technologies of inhabitation" could even aid humans in lending their creativity to weaving those loops further. Through such accumulating degrees of participation, humans gradually learn to incorporate themselves inside the depth of the metamorphic terrain. Just as the land successively becomes more articulate, and just as different places begin to speak more lucidly with different voices, the presence of humans becomes more meaningful, eloquent, and integrated.

It is also implausible to conceptualize indigeneity as a linear progression toward a coveted state of "highest perfection." Indigeneity is, at best, an imperfect competence. The modern tradition cultivates its epistemological optimism on the grounds that reason has gone into exile, and from its isolation, reason is thought to continuously improve the human condition. This phenomenological reading of indigeneity offers a contending possibility: that the competence to live in place indefinitely cannot be won through a succession of reason-derived deductions and then possessed ever after. It requires periodical renewal, rejuvenation, recomposing. It is hard and risky work. It can fail.

It can also succeed. Even very high densities of humans can live alongside large numbers of wild salmon for thousands of years. This truly remarkable feat was already accomplished once in the Pacific Northwest. Why should it not be accomplished again, there or elsewhere?

Once we discard the historic linear connotations of "indigeneity," we can ask: What will it take for contemporary people living right now in each of Earth's regions to articulate their kind of indigeneity, resonant with their epistemological horizon, responsive to the animate powers of their land? Each community, summoned by a place, could channel the flow of their (multi)cultural imagination toward a greater integration with Earth, regardless of the technologies they wield, regardless of their past track record of ecological ignorance, ethnic conflict, human and more-than-human trauma caused or suffered. The land is always striving to sing through us. We are always already being spoken to. We are being enticed into becoming more "geoliterate"—into learning the delicate art of how to *read Earth*. How to *speak Earth*. How to create our niches as storytelling animals inside Earth's larger evolutionary story.[32] Given the now omnipresent magnitude and speed of anthropogenic ecological collapse, this is not merely a matter of poetic speech. Given the inconvenient evidence that the current geological epoch might already be slipping away, such questions become a matter of life and death.

CHAPTER ELEVEN

In the Shadow of the
Standing Reserve

Now I am become Death, the destroyer of worlds.

Bhagavad Gita

S omething stirs in the shadows. The modern technological complex is daring dark agencies: It is estimated that the worldwide fishing fleet is now between two and three times larger than would be necessary to take out present-day catches of fish and other marine creatures. As recently as the 1950s, industrial fishing fleets were concentrated on the North Atlantic and the North Pacific, but since then the fleet has expanded in successive waves in all directions, and now there is not a pocket left anywhere that is beyond its reach.

One axis of expansion was the geographical expansion into lower latitudes: From the North Atlantic and the North Pacific, fleets expanded southward into the Eastern and Western Central Atlantic, onward into the Mediterranean and Black Sea, into the Northeast and Eastern Central Pacific, and later into the Southern and Western Pacific, the Southern Atlantic, and the Indian Ocean. Fleets then expanded vertically into ever greater depths. Bottom trawlers can now to reach into depths of more than a thousand meters. When their large wheels are dragged along the seabed, they stir up plumes of sediment, clouds of muddy water that camouflage the net from the eyes of the targeted fish and other creatures. The noise attracts inquisitive fish, causing them to swim before the net mouth. The trawlers continue moving at roughly the same speed, and the fish continue swimming until they are exhausted enough to slip back into the net and get caught. What happens inside those bottom trawlers is grim: "The nets rip hundreds of tons of animals out of the ocean, squeezing some of them so tightly against the sides of the nets that their eyes bulge and burst out of their skulls. For hours, trapped fish are dragged along the ocean floor with netted rocks, coral, and ocean debris. Many fish's scales are completely ground off. When hauled out of the water, surviving fish undergo excruciating decompression. The intense internal pressure ruptures their swim bladders, pops out their eyes, and pushes their esophagi and stomachs out through their mouths."[1]

Bottom trawling has been nicknamed the clear-cutting of the seas, and Les Watling, professor of zoology at the University of Hawaii, describes it as more destructive than any other human activity in the oceans. Watling says: "Ten years ago, Elliott Norse and I calculated that, each year, world-wide, bottom trawlers drag an area equivalent to twice the lower 48 states."[2] Watling has observed the sediment plumes that result from this bottom trawling through satellite images, which are readily available[3] and, in their own way, oddly artistic, reminiscent of children's art or of chromosomes inside a nucleus: Ovals and circles intercept one another at surprising angles, lines crisscross the green expanse of the ocean. Individual lines in the overall scrawl might be anywhere from 10 to 30 kilometers long, and the entire composition might have a diameter of 70 kilometers or more. But these visible traces show only a fraction of the overall impact of bottom trawlers around the world. There are tens of thousands of them in use, and they are as large as they are indiscriminate: Some trawlers will catch up to 20 pounds of so-called "by-catch" for every pound of targeted fish or shrimp.[4]

As the industrialized nations' fishing fleets were pushing out toward new waters, a pattern became visible in their wake. Landings in the North Atlantic peaked during the 1960s. When they did, fleets had already begun to seek new regions. The fleet pushed into the Northeast and Southeast Atlantic, and landings there peaked during the 1970s. In the Western Central Atlantic, Mediterranean, and Black Sea, landings peaked in the 1980s. The fleet sought further expansion into the Southwest Pacific; the peak came in the 1990s.[5] Across the planet, industrialized civilization has decimated 90 percent of all the large fish. It took only sixty years to get there. It is no longer the stuff of science fiction but rather a matter of common sense: At the current rates, all of them will be gone by the middle of the twenty-first century.

Trespassing Planetary Boundaries

The narrative of human exceptionalism has maneuvered itself into a rather wicked dilemma, one for which its inherent optimism has left it ill-prepared. A structurally immanent irrationality burdens the superrationalistic project of modernity—one that is inherently self-destructive. There are thresholds in the biosphere that should never be crossed, and yet the combination of epistemological optimism and ever more powerful technologies have already pushed beyond some of these thresholds. An influential report by the Stockholm Resilience Centre, published in 2009, identified nine so-called "planetary boundaries."[6] Together, the nine make up a safe planetary operating space within which humans could thrive indefinitely without upsetting the relatively stable conditions of the Holocene epoch, the geological epoch inside which humans have lived for the past ten thousand years. The report suggests that pushing even one of these beyond a critical threshold would be enough to seriously upset the Holocene stable state. But the report also shows that three of those factors have already been pushed beyond a critical threshold: climate change, biodiversity loss, and the nitrogen cycle. This might come across as cold statistics, but it is not. It is evidence that the planetary community of life, including the chemical composition of the atmosphere and of the oceans, has already been altered in fundamental and irreversible ways. There is a growing consensus that the Holocene epoch is already slipping away. An entirely new Earth epoch is now upon us.

This is a moment in Earth's very long story when a disaster has come to pass—one species among the multitudes has become a destructive geological force. The planet is transitioning into a new chapter in its ancient story, a chapter marked by greater uncertainty, as a single species has managed a feat none of the estimated four billion other species who evolved on Earth throughout its life history had ever accomplished: to initiate a wave of drastic biospheric changes, setting the Gaian story on a course likely to be very, very different from anything this planet has seen before.

After several centuries of treating the planet as a *res extensa*, and of thinking of Earth as devoid of any agency other than human rationality, the old optimistic dream of mastering Earth through the exclusive power of reason has morphed into a waking nightmare. While the modern lifeworld is still escalating its impact on the biosphere, Gaia's more menacing aspects are becoming rather articulate. Their ferocity is measured not on human scales but on geological scales—they respond through the word of Gaia.

If their individual powers are difficult to gauge, their combined potency is unknown and unknowable. What does seem clear is that their concerted actions play out at scales and at orders of complexity vastly beyond ratio's capacity to contain, to manage, to control, often even to imagine. Earth's agencies will not be contained. They do not hesitate to bite back. The story of separation is now on a frontal collision course with such frightening agencies as ocean acidification, or the oceanwide collapse of food webs as we know them. It has become a force so powerful that the evolutionary course of this planet is being rewritten for all time to come.

———

Dark forces are stirring. Agencies are rising that will no longer be quieted: At no time in the last 650,000 years has the oceans' basic chemistry changed more rapidly than now. As with anthropogenic climate change, ocean acidification results from pumping excess carbon into the air. The difference between climate change and ocean acidification is that the latter is a far more straightforward and linear process. If more carbon dioxide (CO_2) is dumped into the atmosphere, more of the gas will be absorbed by the oceans, which will consequently become more acidic. The CO_2 that sinks into the oceans will mingle with salt water to form carbonic acid. If you have ever had a sip of Coke or carbonated water, you know what carbonic acid feels like in the mouth. And if you have ever put a baby tooth in carbonated water for two or three weeks, you know why this is a problem.

Ocean acidification has been going on for roughly two hundred years, ever since industrial civilization began treating the atmosphere as a giant waste dump. Throughout those two centuries, over 1 trillion tons of CO_2 have been released into the atmosphere, of which 525 billion tons, or nearly half, were absorbed by the oceans. The oceans absorb 22 million tons of CO_2 every day.[7] Christopher Sabine, one of the researchers who have worked hardest to give the issue broader public attention, has tried to make those numbers more palpable by pointing out that 1 metric ton is roughly equal to a Volkswagen Bug, which means that the oceans absorb twenty-two million Volkswagen Bugs' worth of CO_2 every single day. And the rate of absorption is still rising. "Forty-three percent of [the overall amount] has happened just in the last 20 years," writes Sabine. "And it's growing exponentially. As we start to project out into the future, it really gets scary. And people say, oh, the oceans are huge, we could never really affect the oceans. Well we are. They are changing, and we are measuring it."[8]

When CO_2 is absorbed by the water, it lowers the surface ocean pH. So far, the worldwide pH has dropped by 0.1 unit below the preindustrial state to 8.05. This might sound insignificant, but the relevant detail in the equation is the fact that the pH scale is logarithmic, which means that for every 0.1 drop, the ocean's overall acidity increases by 30 percent. At the projected rates, the pH will have dropped by another 0.3 to 0.4 unit by the end of the century. This means that the overall ocean acidity will have increased anywhere from 120 to 150 percent as compared to preindustrial times. "We, the scientific community, were wrong," writes Dr. Edward Miles, "we had underestimated both the magnitude and the rate. . . . Are we involved in another mass extinction event, for corals, for instance? When such events take place, it takes millions of years for the corals to come back. . . . Are we screwed? Yeah, to a considerable extend. A world of 500 ppm is a world of enormous environmental destruction. We ought to recognize that, and *say* it."[9] As a report by Richard Feely and his colleagues puts it: "The decisions we make now, over the next 50 years, will be felt over hundreds of thousands of years."[10] The message is unambiguous: The changes now under way frame the implications of our present actions in terms not of years, not of generations, but in terms of geological dimensions.

The oceans are home to many creatures who depend on a mineral called calcium carbonate, which is the material they metabolize to form their shells and exoskeletons. When CO_2 enters the water and creates carbonic acid, it will reduce the abundance of carbonate ions in the ambient water, which in turn will mean a net loss of such ions to the crustaceans, mollusks, corals, and echinoderms who use them to form their calcium carbonate skeletons and shells. Up until today, only a fraction of the creatures who need calcium carbonate to secrete shells or skeletons have been researched. So far, nearly all of those who have been researched have been affected by the acidification of their environment.

Take pteropods, or, as they are also known, sea butterflies. In high latitudes, these tiny snails can inhabit the upper ocean column in densities of more than ten thousand individuals per cubic meter, flapping through the water with miniscule winglike lobes. Owing to their sheer abundance, they form an intricate link at the bottom of the ocean food web, an essential part of those largely microscopic, errant drifters known to us as plankton. In other words, pteropods are essential to the entire ocean food web. Research on these little drifters has found that their shells will dissolve under ocean conditions that we are likely to see within a few decades.[11] When these sea butterflies were exposed to the kind of environment they are likely to inhabit by the end of this century, it took only two days before their shells started to dissolve into the seawater. The authors also point out that the major cause of worry is not so much the amount of CO_2 added to the oceans, but the rate at which it is being absorbed. At this rapid speed, the creatures do not have the time to adapt.[12]

As the oceans are becoming more acidic, more habitat is being lost permanently to some of the ocean food web's most abundant and most vital specimen.[13] The impacts of this ripple out through the entire food web, to anyone from krill to shrimp to salmon to blue whales. When juvenile pink salmon migrate from their home rivers to the cold and productive waters of the Arctic and sub-Arctic, they feast on the great abundance of sea butterflies. This is where the salmon, like so many others, come to grow and to mature. This, too, is where the Sitka spruce originate, and the bald eagles, and the wolves, and the grizzlies, and the stories of the Coast Salish, and the full bellies of their children. In a study on the feeding habits of juvenile pink salmon, Janet Armstrong of the University of Washington and her collaborators showed that in some years, the sea butterflies comprised 60 percent of the young salmon's diet.[14] Other work suggests that if sea butterflies decrease by only 10 percent, that could lead to a drop of 20 percent in the body weight of mature pink salmon.

These are what a photographer would call extreme low angle shots, or what a naturalist might call a worm's-eye view. We could also call them a "modern view," and thereby acknowledge that our ability to steer clear of these changes has so far been frustrated, in part, by an intrinsic inability to comprehend their scope. We now know some of the basics. We know that more acid in the oceans will likely result in disturbing consequences. We know that coral reefs might not be able to survive the rapid changes of the next fifty years. We know that no shellfish (crabs, lobsters, shrimp, oysters, clams, squid) might be able to live in the oceans by the time our children walk the beaches with their grandchildren. We know that sea butterflies weave the ocean food web together, and that forty-eight hours spent in the projected conditions of the year 2100 would cause their outer shells to dissolve. We know that salmon and many others will starve without shellfish and zooplankton. We know that many beautiful things, both named and unnamed, will die and not come back. But how does it all hang together? What is the bird's-eye view on this—the Gaian view?

The few larger lines that we can sketch out are not very reassuring. Richard Feely suggests that we must go back about fifty-five million years to get a perspective on the present changes under way in the acidifying oceans. Back then, volcanic eruptions and a disturbance of large submarine gas deposits caused a substantial change in basic ocean chemistry, not unlike today. Massive amounts of carbon dioxide were spewed into the atmosphere and were dissolved back into the oceans. The result was a mass extinction wave that reverberated throughout the ocean food web. Granted, life came back after the mass extinction, and in greater diversity than before. But the last time this happened, fifty-five million years ago, "it took the oceans over one hundred thousand years to recover."[15]

According to a study published by Anthony Barnosky and his colleagues in *Nature* in 2011, a so-called "mass extinction event" occurs when three-quarters or more of all species disappear during a geologically brief time. Five such events—the so-called Big Five—are known from the geological record. Are we now seeing the sixth mass extinction? Strictly speaking, counting all the recorded and estimated species who have already vanished or are currently vanishing through anthropogenic causes, we cannot yet speak of the current wave of losses as another mass extinction event. But the combined inertia of so many human-caused stress factors is driving the Gaian biosphere straight toward just that. "Additional losses of species in the 'endangered' and 'vulnerable' categories could accomplish the sixth mass extinction in just a few centuries," write Barnosky and colleagues. "It may be of particular concern that this extinction trajectory would play out under conditions that resemble the 'perfect storm' that coincided with past mass extinctions: multiple, atypical high-intensity ecological stressors, including rapid, unusual climate change and highly elevated atmospheric CO_2."[16]

Systemic biospheric collapse might seem remote, vague, and difficult to conceptualize on scales relevant to modernity's attention span. But in Gaian terms it is happening very, very fast. The drastic biospheric responses we are now witnessing are occurring just four centuries after moderns began imposing the metaphysics of the machine and of human exceptionalism upon Earth. Nothing in the geological record might come close to the current implosion of the living web. The fastest die-off of the Big Five—the so-called End Permian extinction from about 252 million years ago, also known as the Great Dying—still lasted around 60 thousand years.[17]

Against the horizon of changes such as the oceanwide dying of fishes, ocean acidification, and the planet's sixth mass extinction, we might think of the modern narrative, with its fixation on the ever-new, as being at once very narrow and very large. Its narrowness is in its monocentrism and in its drastically narrowed attention span. Its ability to nurture a concern for the more-than-human community of life is vastly inadequate. It has also not convincingly proven able to inspire effective social practices that nurture far-sightedness or a sense of moderation. On the opposing end, we find the story's sense of boundless entitlement, which expresses itself most force-fully in a technological leverage that is, for all practical purposes, infinite. Like no other narrative before it, modernity has perfected this extreme smallness-in-largeness, that signature blend of near-total self-absorption and limitless power. This narrative's programmatic collapse of the horizon of concern—its monocentrism and its preoccupation with the ever-new—has little to offer when it comes to preparing the citizens of the dawning Anthropocene for changes on the scale we are now learning about.[18]

Magic

The modern technoscientific complex creates an ontological situation that we could describe, with indebtedness to Heidegger, as an obliviousness of Being. In Heidegger's thought, modern science and technology are erected on the fundamental assumption that through them, it will one day be possible to make Being fully explicit and knowable. When that happens, it will be possible to fully control Being. Being will be at humans' disposal as a "standing reserve" that the human animal can direct, manipulate, and recompose at will. Remember Bacon's vision that science and technology would bring about the "fulfillment of the millenarian promise of restored perfection."[19] And Descartes said: "Give me matter and motion and I will construct the universe."[20] This same epistemological optimism echoed later in the words of Karl Marx: "Freedom in this field [of production] can only consist in socialized men, the associated producers, rationally regulating their interchange with nature, bringing it under their common control, instead of being ruled by it as the blind forces of nature."[21]

Boundless epistemological optimism has gone hand in hand with an equally boundless utilitarian optimism: Because the rational intellect can supposedly know it all, it should be possible to build technology that translates this perceived omniscience into omnipotence. This rationalistic ideal has sought to assume firmer and firmer control of this vast Earthen sphere. Human rationality has laid tightening webs of infrastructure across the land, harnessed Earth's rivers, unleashed ancient fires buried in its mantle, cast the forests, grasslands, oceans, all our animal and plant companions, and eventually even the very atmosphere under its management schemes. As late as 1959, C. P. Snow wrote that science has its own "automative corrective."[22] This was only three years before *Silent Spring* empirically showed the lethal outcome of intemperate rationality let loose upon the land. Although there is more empirical evidence now than ever that the biosphere is shuddering with wave after successive wave of systemic collapse, there remains a sense in which the cycle of trauma and loss can only be broken by crafting more powerful technology and exerting more control on the land, until, in a hypothetical future, we will at last catch up with, bypass, and contain those rippling waves of losses. There remains a sense that one still ought to extend rationality's reach, that one ought to clear-cut an even grander view from nowhere, for from there it will one day be possible to restore that lost perfection. And then, to apply that ingenuity so as to orchestrate Earth indefinitely. There remains a sense in which rationality itself is invincible, even when the living world around it is suffering.

———

The conflation of knowledge with applicability is older than the modern story, and it has left some rather instructive track marks in our language. Consider for a moment the etymological trajectory of the word *engine*, a telling symbol for the technical aspect of anthropogenic ecocide.

In 1393, the poet John Gower wrote in his *Confessio Amantis*: "Tho wommen were of great engyn."[23] Later, in the fifteenth century, Henry Benjamin Wheatley wrote in his *Merlin; or, The early history of King Arthur*, "I am the sone of the enmy that begiled my moder with engyn."[24] A century later, Sir Thomas More wrote in *The historie of the pitifull life, and unfortunate death of Edward the fifth, and the then duke of Yorke his brother, with the tyrannical government of usurping Richard the third*: "By what crafty engin he first attempted his ungracious purpose."[25] Another sixteenth-century writer, James Farewell, wrote: "And after long and tedious ranging, By help of Mathematick Engine, A Setting-pole the cunning Rogues Brought from the Fleet to leap the Bogs."[26]

These are instances of the oldest meaning of a word whose roots reach deep into the Indo-European language family. In classical Latin, the root *ingenium* means "natural disposition, inherent quality or character, mental powers, natural abilities, talent, intellect, cleverness, skill." In twelfth-century Old French, the root *engien* means "inborn talent, intelligence, or wit." In twelfth-century Anglo-Norman, the word was even used to speak of "magic power."[27] But early on, these various connotations around the common theme of ingenuity also began to be associated with tools and gadgets. "Engine," in its many forms, became associated with warfare (Milton in *Paradise Lost*: "Whereof to found thir Engins and thir Balls"), with instruments of torture (Shakespeare in *King Lear*: "That like an engine wrencht my frame of nature from the fixt place."), and also with fishing devices (from the eighteenth-century *Sportsman's Dictionary*: "The little stick may have one end in the notch T of your tricker, and the other end in the hole X, and then is your trap or engine set right as it should be.").[28]

The idea that tools reflect their makers' ingenuity appears to be an old and familiar notion to the Indo-European linguistic community, one that predates the early modern philosophers and the early canonizers of scientific thought. Modern thinkers did not so much invent the conflation of human ingenuity with technology but codify this conflation into a stringent method and gave credence to a sentiment widespread throughout much of the Eurasian continent. It makes sense, then, that the pioneering engineers who helped build the industrial age would have identified their craft and their machines with "great cleverness," and perceived of their sudden and unprecedented boost in power as the closest thing to magic humans had ever seen.

The notion that the rational intellect can predict, contain, and manage technology's every impact on the biosphere is yet another variation of that now familiar theme, namely the belief that mind exists in a bodiless and purely abstract space somewhere outside the biosphere. Planetary ecocide and technology's still-accelerating power are evidence that we have not yet absorbed, collectively, the mind's categorical entanglement with the biosphere in a technological imagination that might truly call itself "ecological." We have not yet recognized that ignorance will inevitably increase proportional to technology's accumulated leverage on the fabric of life, precisely because mind is inside the living web.

This is the curious paradox of modern technology, and the inherent problem with the die-hard habit of conflating ingenuity with applicability: While modern technology so clearly articulates and perpetuates the early modern metaphysics of control, dominance, and separation—while it still chases the Baconian dream of creating a kingdom of perfection— the modern technological imagination actually leads to us being less in charge, to there being less beauty, more suffering and unpredictability and fearfulness in the world. As modern rationality translates its dream of omniscience into tools of omnipotence, it ignores the truth of our entanglement with the sphere of life. A technological imagination that is truly ecological would integrate this fundamental and comprehensive entanglement into technology.

We can assume that no one ever *wanted* the oceans to turn acidic, but the fact is that the chemical balance in the oceans began to shift precisely at the time when combustion engines were first used on a mass scale. The absorption of CO_2 by the oceans and the massive disruptions to food webs now under way took everyone by surprise when the reality of it recently started dawning in our collective consciousness. It took roughly two hundred years between the creation of the engines and the time when researchers first became aware of the problem. Whether reason will now be able to contain the forces it has unleashed, is not certain.

Mostly, it seems that we are discovering how tremendously little we actually know about ocean acidification in the industrial age. Our enormous technological leverage over the oceans has led to a proportionally monumental increase in our ignorance about the waters and the creatures that inhabit them. One recent study, for example, admits that "most of the elevated CO_2 response studies on marine biota . . . have been short-term experiments that range from hours to weeks," but that longer term exposure "*may* have complex effects" and "*could* induce possible adaptations that are not observed in short-term experiments."[29] In short, it doesn't tell us much of anything, at least not anything useful. Studies on calcite-forming ocean algae and phytoplankton "have covered only four of approximately 250 to 500"—that is, between 0.08 and 1.5 percent—of the living species.[30] The reports seem primarily to serve as lists of all that is still unknown, including arguably the most important piece: "Little information is available . . . for potential ecosystem-level consequences" that will result from carbon dioxide levels "that are predicted to occur over the next 100 years."[31] As the authors of another study bluntly observe: "The effects of chronic exposure to increased CO_2 on calcifiers, as well as the long-term implications of reduced calcification rates within individual species and their ecological communities, are unknown."[32]

No one really has a clear understanding of what is being unleashed, not even those who have made it a career to study this. This wicked dilemma is a direct consequence of the epistemological optimism that has been inherent in the modern story since its beginning. The decisions we make now will be felt over hundreds of thousands of years.

Nature as *Res Extensa*

For Heidegger, the question of modern technology is inseparable from the way the modern narrative tradition has related to what we are used to calling nature. The Latin word *natura* ("birth," from Latin *nasci*, "to be born") was a translation of an earlier Greek word, *phusis*. *Phusis* is derived from *phýein*, a temporal word whose basic meaning is "emergence." What we call *nature* was originally encountered as something that arises from within itself, as the world's tendency for self-emergence. Heidegger's thought on Being and on Earth reaches back to the sense of this Greek root, *phusis*, combining it also with the concealing, withdrawing aspect of phenomena discussed earlier. As the Heidegger scholar Bruce Foltz puts it, "In Heidegger's thinking proper, the Earth is not only that in which plants take root and upon which houses are built but also the human body, the sound of a word or the script of a text, the bronze or clay that upholds a sculpted surface. In each case, the Earth is what bears and gives rise to what comes to light by remaining intrinsically dark itself."[33]

What Heidegger observed in his study of early Greek thought is foremost an animistic, embodied sense of Being. Practically none of these animistic connotations have remained resonant in the Latin-derived *nature*. The word *nature* no longer reverberates with the land's immanent vitality, as that vitality expresses itself uniquely through each creature and each phenomenon in the landscape, including rivers, including wolves, and including also the human animal. The word *phusis* once voiced an experience of Being that seethed with agency, that was never merely "at hand," but poured itself forth as uniquely in a juniper tree as in an avalanche, delivering itself intrinsically into the world while never fully standing out in the open, and always reserving the possibility to withdraw again from unconcealment. By having this word, *phusis*, early Greeks could recognize in everyday speech the many other voices or actors in the land, meek or grand, patient or explosive, moderate or frightening. They could recognize the land's speech right there beneath their feet and feel themselves embedded in a throbbing crowd of agencies that far exceeded the human animal's local and fleeting capacity for rational thought. To stand with their two feet planted firmly inside such an articulate landscape would have been to be immersed in an ineffable, burgeoning creativity that continuously oozes forth from an ineffable abyss, lingering in the open for a while and then returning to nourish that abyss again. As Thomas Berry writes: "Because creatures in the universe do not come from some place outside it, we can only think of the universe as a place where qualities that will one day bloom, are for the present hidden as dimensions of emptiness . . . qualities arise from emptiness, from the latent, hidden nothingness of being."[34]

Nature, in the Western metaphysical tradition, gradually became less animated, until, at the advent of modernity, it was interpreted as machine-like, exposed in the open where reason could know it completely. It was interpreted as opposite to humans, who had forgotten by then that their own species name had once planted them firmly within this Earth, that they were earthlings among a multitude of other earthlings. Nowadays, the word *nature* is but another marker of separation. The best use for it is to exclaim that "humans, too, are part of nature." If this sounds hollow, this is not only because the rich semantic resonance of our own species name has become obscured, but also because *nature* has become thoroughly de-peopled. No voices speak to us through the word. It is a semantic corpse, deader than the stricken but still-pulsating word *human* (struggling, as it does, to reintroduce itself as earthling). *Nature* does not resonate with the primordial connotations that humans, too, like estuary alders, like ravens, like granite cliffs, like the drifting continent-elders, participate in Earth's self-birthing, that humans, too, are Earth gathering itself in moments of heightened meaningfulness and subjec-tivity, that our lives, too, unfurl, in Thoreau's words, in "the bloom of the present moment,"[35] and that this bloom is peopled by legions of other-than-human agencies, each of whom has agendas indifferent to what we are up to and who will not be contained by the rational intellect.

A Story Charged with Immense Mythic Power

Earth's ontological status, in a technological age, is to be a resource, to become one more component of the world-spanning technological inventory. It is to be forced to be what Heidegger calls a "standing reserve." Each aspect of Earth is presumed to be available at all times. Nothing is allowed to remain in obscurity. Any mountain, forest, river, or species is forced to be a deliverer of raw material or raw energy whenever the technological framework requires it. Such an ontological situation is one where Being is thought to be fully exposed, and forced to endure relentlessly in the open. Foltz calls this revealing a "regulatory attack."[36]

When this living membrane that is Gaia, this breathing body, is cast into radiant exposure as an assemblage of inert stuff—a *res extensa*—what becomes concealed is the fact that it *is* a body, that it has the tendency to hold aspects of itself in concealment, that Earth is organic and shapeshifting, that the round planetary body draws the human animal back again into its far vaster, storied metamorphoses, that the vast array of animate Earth-actants composes the Gaian whole in loops upon loops of mutual participation, and that the biosphere acts in ways so complex that the rational intellect will never be able to fully predict or manage. What becomes concealed is the fact that because we are mindful bodies participant with this vaster body, we cannot hope to conceptually bypass, dodge, or overtake this planetary presence, within whose curiously unstable atmosphere we are bound. Human rationality, too, is an Earth characteristic. We remain firmly within. In revealing Earth as a static object that is constantly available, bare and visible, modern technology conceals this participatory, animistic dimension of Being and our organic involvement with it.

There is surely something obscene about this, something crude and offensive. And there is also a certain irony there. As Earth is cast into rigid categories, as it is subjected in its entirety to expectations of utility, the living world is stripped of its many agencies, interpreted as a machine made up of replaceable component parts. But then humans, like everything else, are also replaceable. They, too, become spare parts in a world-spanning metaphysics that knows no living creatures, no sentient bodies, no delicately reciprocal participation. Heidegger's fundamental critique of the modern technological framework was that it presents the culmination of modern metaphysics, drawing every aspect of our relationship with Being into its gravitational force field. The mind-world divide assaults not only other-than-human life but also our own feelings, intuition, or the senses. This culmination of metaphysics will turn against those aspects of ourselves that are habitually dubious to the rational intellect. The human, too, is being further partitioned through modern technology.[37]

As the planetary topsoil is growing thinner, as the oceans are literally being fished empty, as the long-term ramifications of ocean acidification are shown to be fully unpredictable, as death tolls continue to rise throughout the biosphere, even as the atmosphere is heating up, the Norwegian salmon industry is envisioning a fivefold growth within the next human generation.[38] This seems like a paradox at first. But why, we might ask, does the vision of a fivefold production increase not only seem attractive but actually possible, from within a certain narrative?

At least part of the answer is that technology's ontological authority is so ubiquitous. The fact that "technology works," and that gadgets are still becoming ever more exquisite and formidable (within narrow, rationalistic, utilitarian limits), gives technology considerable perceptual persuasiveness. Seen from within the narrative of modernity, it is still difficult to resist the medieval temptation to conflate utility with magic, and its particularly modern articulation as boundless epistemological optimism. The simple and obvious fact that technology works so well charges the modern metaphysics with immense mythic power.

It is thanks to this mythic power that epistemological optimism remains structurally immanent in modernity. Modern megatechnology perpetuates itself through a hypereffective feedback system. Technology creates the kind of short-term successes that in turn justify and strengthen a cosmology of human dominance, which in turn generates and reinforces its own set of inherent value systems and sets of cultural assumptions. The omnipresence and sophistication of technological spaces create another feedback loop, giving modern humans the impression that the world really is created by them, and for them, and that human ingenuity really is the measure of all things.[39] Wherever you look, you can see: Production is still growing. Salmon are so abundant that they are practically available in every grocery store, anywhere, anytime. This is the nature of being-in-narrative: From within a particular narrative, it is almost possible to believe that all is well.

CHAPTER TWELVE

The Salmon Fairytale

*Modernizers are extraordinarily good at freeing
themselves from the shackles of their archaic, provincial, stuffy, local, territorial past, but when the
time comes to designate the new localities, the new
territories, the new provinces, the new narrow
networks toward which they are migrating, they
content themselves with utopia, with hype and
great movements of the chest as if they were preparing themselves to breathe the thin intoxicating
air of globalization. No wonder: they never paid
any attention to* where *they were heading, obsessed
as they were to* escape *from attachments to the old
land. . . . At the epoch of the Anthropocene, the great
narrative of emancipation has made us totally
helpless at finding our way to where we belong.*

BRUNO LATOUR

O nce upon a time there was a small kingdom in the far north called
Norway, the "Way to the North." The kingdom had been blessed
with so much natural beauty that it seemed almost as if from another
world. Fjords cut deep into the mountains to the west. Anywhere you
looked, waterfalls threw themselves down into deep gorges. Glaciers
spread their huge, white tongues into misty valleys of lush evergreen forests, where moose roamed, along with lynx, bears, and red deer. There were
places where the sun never rose in winter and never set in summer. There
were places where one could not cross a mountain pass without stepping

on dozens of lemmings whose population had just exploded again in spring, curiously and wondrously. Anywhere across the kingdom, aurora borealis might suddenly materialize in the starry night sky, performing her otherworldly magic. There were places where one could move in a single day through winter, spring, and summer, beginning the morning on skis way up on a mountain plateau (watching a herd of grazing wild reindeer), spending midday wandering through flowering meadows of coltsfoot and anemones, and ending the day with a refreshing plunge, naked, into the mild waters of the North Atlantic (while a family of killer whales might be passing by on its way into the next fjord).

This small kingdom had suffered a long history of oppression by larger Scandinavian powers. The Danes had ruled over it for three hundred years until the Norwegians rose up in 1814 against the old oppressor and drafted their own constitution. But the Swedes denied Norwegians full independence for another ninety years until Norway dissolved its union with Sweden in 1905 and jump-started a new lineage of kings. From then on, it seemed clear to Norwegians that the young century would be theirs. The twentieth century would be a time for soul-searching, for completing the nation building that had begun in the nineteenth century. It would be a time for finding out where they belong.

As it turns out, modern Norwegian identity was shaped by two stories that developed around the immense riches of resources brought up from the oceans. In the 1960s, Norwegians struck oil, and the notorious story of Norway's conflict-ridden entanglement with fossil fuels has been known, ever since, as the *oljeeventyret*, or Oil Fairytale. The salmon industry emerged nearly simultaneously. Not surprisingly, this other success story has since become known as the *lakseeventyret*, or Salmon Fairytale. Salmon, like oil, are considered a unique and spectacular success, an exceptional modern marker of what it means to be Norwegian.

Lisbeth Berg-Hansen, minister of fisheries during the Stoltenberg administration and also a major shareholder in a family business that ran salmon production units, once called the salmon feedlot industry "part of the Norwegian identity."[1] Prime Minister Jens Stoltenberg himself repeatedly let journalists depict him advertising farmed salmon during his diplomatic travels abroad. Members of the royal family have also released press images from travels abroad wearing aprons and holding stainless steel knives, with salmon fillets laid out in front of them ready to be sliced. Stoltenberg's successor as prime minister, Erna Solberg, is continuing this close alliance between politics and the feedlot industry. In March of 2015, she announced on national radio that "salmon is for Norway what Ikea and Hennes & Mauritz are for Sweden—our prime international brand. It is the number one product people abroad associate with our nation."[2]

But there is something cynical about these stories: In the cases of salmon and of oil, national trauma is being recruited to make profit at the expense of others. In the salmon industry, the "others" are the salmon themselves, and those communities of beings who inhabit the seas around the pens. In the case of oil, the "others" are all those who must face the short-term and long-term consequences of the accumulating CO_2 in the atmosphere. In subtle ways, the salmon industry, like its oil counterpart, works to anchor itself emotionally in the collective psyche as something worth defending, something to be proud of, even when its internal moral dilemmas are so numerous. The Norwegian salmon industry suggests that it is unproblematic to create a national epos around the systematic exploitation of hundreds of millions of sentient creatures. In the process, citizens become redefined as consumers, salmon become redefined as commodities, and community becomes redefined as a collection of alienated, insular beings, aggregates in a larger technocratic framework. Heidegger's thought was prescient.

Starving Rationality

While all this is done in the name of rationality, the practices and the story into which they are embedded are built on unexamined dogmas that are not rational but ideological. They include the followings ideas:

1. It is reasonable to anticipate exponential growth curves for the salmon industry even in the face of eroding soils worldwide, the comprehensive collapse of fish stocks, the acidification of oceans, or other systematic ecological breakdowns.

2. It is possible to predict and contain all short- and long-term effects on all habitats that are impacted by the industry's practices, both local and far away.
3. This is possible because the world is composed of no agencies other than reason-driven humans. There are no tipping points, no waves of action, no feedback loops making their claims on human social contracts.
4. It is physically possible to build economic practices around the metaphor of the world-as-machine and sustain such practices indefinitely.
5. Humans are able to flourish even under the most reductionist and utilitarian interpretation of what it means to be human, where they become defined as consumers and all else is reduced to raw material.
6. Salmon can be thought of in isolation from the open land and seas wherein they have evolved.
7. Reason alone is able to disclose the world truthfully.

A central problem is that of rationality itself. There is an assumption that it is possible to extract rationality out of the concrete context of what it takes to live here inside this living Earth community. The French philosopher Bruno Latour has observed that rationality, comprehensively understood, needs to be embedded fully into the envelopes that are essential for the furthering of life as such. "Any thought, any concept, any project that ends up ignoring the necessity of the fragile envelopes that make existence possible is a *contradictio in terminis*," says Latour. "Or, rather, a contradiction in architecture and in design: it is unsustainable; it does not have the atmospheric, the climatic conditions that could make it livable. Trying to live in such a place would be like trying to save all your precious data to the cloud without investing in computer farms and cooling towers. If you still wish to use the words 'rational' and 'rationalism,' fine, but then also do the work of designing the fully furnished spaces where those who are supposed to inhabit them may breathe, survive and reproduce."[3]

This is what the salmon fairytale fails to do. Like the salmon, the humans in the discourse are "not all there." The discourse is taking place in perceived isolation from the dynamic complexities that co-constitute the biosphere, including our own bodies, as if it were possible for *Homo rationalis* to seek refuge inside a soap bubble even as a perfect storm of ecological collapse is whipping the nearby treetops. Salmon as spare parts; salmon as extractable from ecological processes; salmon as isolated from Gaia's gift-giving housekeeping; humans as disembodied souls unaffected by the destruction of their atmospheric envelopes—this story from nowhere never asks how and where exactly humans are supposed to live. We are also left to guess how and where rationality is supposed to be nourished when so many of the envelopes that sustain it are unravelling.[4]

There is an undeniable genius to the symbolic consciousness that gives rise to modern technology. But time and again, if treated carelessly, it will show its dark side. The philosopher and historian of ideas John Gray describes the uncritical belief in progress and rapid technological advancement as the new religion of our age. But, he says, no matter how hard we make believe, progress remains a very stubborn illusion.[5] There is no linear, quasi-evolutionary march toward moral or political maturity. The turnover of modern technology might continue to accelerate, and ever more potent gadgets might continue to be churned out, but "unlike science, ethics and politics are not activities in which what is learnt in one generation can be passed on to an indefinite number of future generations. Like the arts, they are practical skills and they are easily lost."[6]

This is the very insight we encountered in the Klallam canon, in the way they crafted an indigeneity that is alert to the possibility of becoming too proud, too enamored with the thinking mind's infinite possibilities. We saw that the kind of collective maturity it takes to live attentively in place, indefinitely, must be reclaimed and rejuvenated again and again, lest corruption, forgetfulness, and alienation corrode it. Gray throws one more observation into the bargain: It is possible to be awestruck at the boundlessly snowballing sophistication of reason, while at the same time allowing human life to grow increasingly more "savage and irrational."[7]

Thought that wanders too far from the wellspring of our embodied participation with the land will be affected gradually. It will begin to project upon the living Earth the hardened images of an imagination further and further astray from the tangible world that feeds us with metaphors, perspective, and elasticity. As thought hardens and becomes less creative, the living Earth will be depleted in a tragic and deadly feedback loop. Rationality cut off from its source will suffer in the long run. Rationality itself depends for its flourishing on healthy bodies, healthy streams, a properly balanced atmospheric composition, and oceans full of life, diversity, and complexity.

———

The kind of rhetoric widespread in the salmon industry and its lobby might already signal a crisis of reason. In a world where salmon are driven from rivers, from folk stories, dreams, and the oceans, and where they are plasticized to conform with the ideology of human supremacy, humans might be starving themselves more and more even as production peaks year after year. I argued earlier that without the nourishment of wild, uncontained salmon, the landscape of our imagination becomes more barren. Perhaps the self-centered narrative that the salmon industry and its lobbyists are enacting is just that—a failure of the imagination. The hard-line insistence that only more of the same will ensure human flourishing—more separation, more manipulation, more reduction, more growth, more control, even when empirical evidence suggests strong, causal bonds between each of those ambitions and planetary ecocide—might be an expression of a wounded, afflicted rationality. It is as if the collective mind is trapped in a vicious cycle where the monoculture of thought leads to impoverished rivers, which in turn are less able to feed back to the storytelling animal what it needs to think resourcefully, creatively, flexibly.

Consequently, the storytelling animal seeks refuge even further "in story." Reason withdraws even further into exile. But the collective dream of a fivefold production growth in the Norwegian salmon feedlot industry during the same time frame predicted for the near-total collapse of ocean fisheries is an example of extreme cognitive dissonance. When such stereotypical patterns are repeated often enough, they perhaps become a psychic addiction. One might not be able to imagine a way out of the vicious cycle, but one does know about quick fixes. Highlighting exponential growth curves is an effective way to make sure the sense of loss will never have to be made conscious. Exercising more control over other-than-human creatures is an efficient way to obscure the dawning understanding that the dream of control is slipping ever more vigorously out of reach, now that the weather is becoming wilder. Even in the face of the dramatic developments we now are witnessing, the story's gravitational pull seems irresistible.

The industry is right about one important point. We do not want to starve. We want abundance and good food. We want to flourish as human beings. The industry is just rather misguided in its claims as to what it might take to get there.

What might it take to get there? What could be a philosophically satisfactory response to this question? I suggest cultivating our response through the methodological practice I have been exploring in these pages: alternating between stories, moving from one narrative context to another so as to gain a critical perspective that would remain inaccessible if we remained too firmly "in-narrative." The ex-centric look gives an altered angle on the seemingly familiar. Questions can arise that might otherwise remain unspeakable, unthinkable.

The Work of the Gift

This is where the Coast Salish story of Salmon Boy might offer a promising alternative narrative, so unfamiliar to us moderns in its economic and ontological understanding of the relationship between humans and salmon. The central economic vision of Salmon Boy lies in the expectation that the bones be returned to the river. The returning of the bones is a nonnegotiable term of the contract between humans and salmon (known in Klallam cosmology as the "original agreement"). This is a complex act, resonant both with symbolic and practical implications. It is a token of respect for the individual who has been consumed, invoking an appreciation for the creature who gave her life to feed others. The act also symbolizes the return of the gift necessary to close the cycle of mutual obligations between humans and salmon; it signifies a cyclical economic bond not between individuals but between nations. This bond continues only so long as it is confirmed over and over again. Returning the bones is quite literally an act of thanksgiving.

The gift of the bones also explicitly acknowledges the community aspect so vital to receiving nourishment from the salmon: Humans may take their share, but they must not forget that others, too, partake in the community feast. Salmon do not have an exclusive contract with humans alone; they have other obligations to fulfill to other hungry mouths and roots. These layers of meaning resonate in the returning of the bones: the respect for the individual salmon, the investment in cyclical returns, the explicit acknowledgment of the more-than-human context within which human-salmon relations unfold. Finally, these three cross-referencing subtexts congregate in one unified principle that underlies the act: *reciprocal gift-giving*. Enacting this principle might be the most significant aspect of the returning of the bones. It might even be fair to say that the story of Salmon Boy suggests that the principle of reciprocal gift-giving might be the most vital precondition for any flourishing human-salmon relationship.

As metaphor and economic model, "the gift" has already proven to be immensely practical, able to spawn flourishing and sustainable human-salmon economies. Prior to the arrival of modernizers at the Pacific Rim, it was the economic model throughout the region, enabling a coevolution of salmon and humans that spanned several thousand years and that resulted not only in seasonally abundant coastal streams but in a richly diverse lacework of human cultures. Both as metaphor and economic model, it must have contributed to the periodic renewal of collective maturity throughout the region.

Salmon are gift-givers par excellence. They gift the land with their remarkable annual inflow of nutrients, a tale told so eloquently by ecologists. They also make more subtle gifts, gifts that are more difficult to recognize through the quantitative, rationalistic screen of mainstream economics: Salmon engage in the hard and persistent work of community building; they help evoke a sense of place; they aid storytelling apes in situating themselves more resourcefully within a particular landscape; they mentor humans in crafting a mindscape that resonates with the land; they bring flow and structure and elasticity to human thought. Salmon feed not only the flesh of bodies but the flesh of the imagination. They gift people with resourcefulness, creativity, rationality.

The gift metaphor itself is a fine illustration. It was likely through observing the return of the salmon year after year that the metaphor of the gift found its way into the structure of human reason in the first place, and from there, into human social structures. Seeing this incredible abundance of food brought from the oceans over and over again almost certainly instilled a strong sense of having been gifted with something very precious. Observing this periodic gift must have evoked a certain economic insight: namely, that the most direct and practical way to keep the gift going is to reciprocate the gift! Their full-bodied, cumulative attention to the lives of salmon would likely have enabled these humans to attune their thinking minds to the needs of the land itself. It would have gradually birthed the metaphor of the "gift cycle," in all its rich overtones.

Fritjof Capra and Pier Luisi, in *The Systems View of Life*, write that "the very structure of reason arises from our bodies."[8] We might add that the structure of reason arises from our bodies' fluid participation in the more-than-human world. The metaphor of the gift illustrates the embodiedness of reason itself. The gift, it seems, cannot be accurately understood as an exotic symbol from a distant culture, or from an alleged mythological past. It might be more accurately understood as a prototype metaphor: a metaphor that strives to surface in the collective psyche over and over again—in new shapes and contexts—as our own kind encounters salmon, and as we struggle to live well and sustainably in their presence.

Salmon feed not only the flesh of bodies but also of minds. Encountering them in their awe-inspiring otherness radiates fresh insights and ideas, gifts that make reason more dynamic, more integrated, and more attuned to the metamorphic land. Encountering salmon aids reason in becoming resonant with specific places and uniquely local seasonal patterns. It assists reason in slipping back inside the land, lest reason find itself (as it does every now and then) tempted to wander astray in soliloquy.

And because salmon feed not only humans but so many others, they also spawn far more opportunities for reason to be nourished and enriched: Every spruce, rosehip, lady bug, green woodpecker, ant nation, or wolf pack brings more integration, more diversity, more dynamism, more beauty, more surprises. Each added subjectivity and every unexpected association presents an opportunity for us to be seized by awe, to be captivated by otherness, to find ourselves abandoned to wonder. That is how the embodied mind can become less fixated, less neurotic, less self-centered, as well as richer, craftier, more cohesive, more well-fed, and indeed more rational. More fully present.

Feedlot salmon cannot offer any of these gifts to us. Shrink-wrapped into identical frozen units in the supermarket and available year-round, they can feed only the most rudimentary and crudest of hungers. Even when their packaging flashes with patriotic tropes, with flags and folk patterns and images of lush fjords, these salmon are hardly able to claim anything other than passing attention. Although the industry does try very hard to evoke a sense of belonging, feedlot salmon can only barely scratch the surface of the imagination. They are unable to slip into speech, into the patterning of our seasonal behavior, into dance and story and song lyrics, into social contracts that mediate between them and ourselves. All this is the work of the gift. It is work that wild, uncontained salmon strive to do.

By suggesting an economy of reciprocal obligations, the metaphor of the gift also makes claims to us humans: Not only do we receive from the salmon the gift of body and mind. Not only do we benefit from their bounty. The gift metaphor is cyclical: We are being reminded of our obligation to give something in return. This is quite inconceivable within the monocentric narrative enacted by the feedlot industry.

Participatory Tools

Even at the most basic level of grammar, the feedlot industry negates the obligation to make return gifts to the living creatures it exploits: The grammatical structure of *we-produce-salmon* (subject-verb-object) is so intrinsic to the feedlot discourse that it is practically unquestioned. The banality of the grammatical structure nearly obliterates the act of exploitation. Salmon are objectified through an ordinary and apparently harmless speech act. Yet, as a result, they are stripped of their agency. Their ontological status becomes reduced to that fixed by the grammatical structure: a *res extensa*, an object, passively receiving the commands of the subject. This elusive but complete inversion of agency runs directly counter to the explicit acknowledgment of the fishes' agency in Klallam cosmology. There, the fish cannot be objects or commodities. They are affiliates or contractual peers, allies in a more-than-human league of nations. Not only do they as a collective possess sovereignty equal to the sovereignty of the human collective—each individual salmon also possesses as much personhood as does every human.

It has been so necessary to write this phenomenology of story precisely because being-in-narrative often has the appearance of being so banal. Remember that the phenomena in Heidegger's thinking withdraw into concealment, and there linger, in plain view and yet enigmatically out of sight. Having explored the phenomenology of our modern narrative and the way in which it conceals itself—but often conceals itself in plain view—we are now able to read the subtle subtext of a seemingly inconsequential expression such as *we-produce-salmon*. What lies concealed in the sentence, in plain view, is just the reiteration of that certain narrative of human supremacy.

We have gathered evidence to argue that in many ways, humans are still being asked to craft narratives that recognize salmon in their thickly textured agency. This agency is not to be confused with the narrow anthropological definition of agency we encountered earlier. Salmon are not reducible to the question "to eat or not to eat." Their agency cannot be recognized in its full complexity if we leave the underlying premise of their commodification unquestioned. Their Being cannot be fully encountered within a metaphysics of constant availability, within which salmon are defined as quantifiable units of flesh. A salmon is Earth, birthing itself through her, gathering itself into a pulsating center of sensuality and intelligence. She is the sun's gift of energy. She is the stratified, accumulated wisdom of all earlier generations, distant ancestors who have gifted her with such magnificent accomplishments as cell symbiosis, oxygenic respiration, eyesight, the sensitivity of skin to the touch of another, or the strange ability to sense Earth's magnetic pull inside her own body. Through her, all that earlier living draws itself into a creative and fully fresh expression of life, a unique and deeply personal sense of inwardness, or subjectivity. In this, salmon possesses an agency that is nondualistic, relational, and at once collective and individual. She is a living, intelligent being who partakes actively and creatively in the composition of her planet's larger animate body. She *is* that body, at once ancient and vigorously of the presence, sensing itself through the keenly evolved sensitivity that is characteristic of her kin.

To speak in ways that allow for such resonances; to resist the temptation of objectifying the Other even at the level of grammar; to hold perception open to the more-than-human agencies who populate this biosphere alongside the rational intellect—these are challenges that the feedlot industry fails to even recognize.

This calls to mind our earlier emphasis on the importance of poetic speech, again in the tradition of Heidegger's thought.[9] Working actively with the poetic dimension of narratives implies that we cultivate an awareness for the speech of things, and that we allow into our discourses the voices of those who articulate themselves in vastly different ways. The poetic strangeness of some parts of this work has been a necessary facet of this. I have trained my gaze for the possibility that what reveals itself as the most "ordinary" might sometimes be concealing a strangeness of profounder consequences. And vice versa: What might seem outlandish and perplexing from within the habitual gaze of the dominant narrative might eventually turn out to make more sense, to be truer.

Audrey Lorde's striking observation remains valid: Poetry is not a luxury. Poetry remains an indispensable and irreplaceable epistemic path to insight. If salmon are to thrive, and if the storytelling ape is to thrive alongside the fish, then, it seems, this would involve an effort to *speak differently.*

In concrete practice, we see further evidence of why the anthropocentric and ratiocentric narrative enacted by the feedlot industry is currently incompatible with the principle of reciprocal gift-giving: The industry-wide trend is toward more rigid measures of containment, toward more manipulation, toward less reciprocal participation. Dead salmon are buried under layers of plastic, an act that directly renounces the responsibility to gift back the bones to the land. The salmon industry manipulates the seasons so as to hasten the rearing process, an act that unbends the cyclical nature of seasonal return and makes salmon constantly available, again robbing salmon of the possibility to gift themselves, for the act of gifting requires a preceding period of privation. Rearing units are being moved inland, an act that cuts the salmon's ties to the ocean and to watershed habitats and thereby deprives them of the possibility to feed other-than-human beings, bringing them into the exclusive domain of humans.

Such technical practices are conceived from narrative, are coextensive with narrative, and feed back into narrative, strengthening it, gradually making it appear more "normal." To the degree that these practices become more widespread and more internalized, they increasingly unbend the principle of reciprocal gift-giving until its multilayered, cyclical geometry disintegrates among the anthropocentric and linear geometry of the feedlot industry. If the economy of the gift is indispensable to any form of indefinitely flourishing human-salmon relationship, such practices are unacceptable.

While it might be in the interest of maximum profit to create a steady stream of salmon flesh that never ceases and never fluctuates, and while this maximization might be sold to Norwegians as a marker of their identity, this flattening out is in fact deeply troubling: When anyone can buy salmon anywhere, anytime, there is no longer any reason to celebrate the gift. The salmon's overexposure makes them invisible. Eating becomes less meaningful. The encounter with the Other becomes pale, boring, unengaging. It is safe to assume that storytelling humans would never have invested so much creative energy into this particular encounter with the Other had it not been for the fact that salmon have always, until very recently, evaded our constant gaze. Our pancultural heritage would be that much poorer, deprived of that many stories and metaphors, had salmon not continued to draw the human animal into the ever-charged dialectics of absence and return, of hope and gratitude, of longing and fulfilment.

A gift economy produces neither "waste" nor other "externalities." It is charged with strong incentives to reintegrate every aspect of an interaction back into (food) cycles. Freeman House once described the indigenous rites and regulations in the human-salmon relationship on the Klamath River in Northern California. He observed that the tools and technology used to pull fish from the river became part of the ritual and were themselves returned to the river following the catch. The weirs, after they had served for a strictly confined time as salmon traps, were dismantled and handed over again to the river. The parallel to the story of Salmon Boy is striking. Here a return gift is also made, a gift critical to sustaining the contract of mutual benefits. The weirs were not just "catching devices." Rather, they were conceived relationally from the very outset: Catching (or rather: receiving) salmon was only one aspect of the Being of the weir. The weir was not only a tool for taking; it also became, at a designated stage in its life cycle, a tool for gifting back!

These bipolar qualities of the same technology were complementary: One was inconceivable without the other, the act of taking conceptually implausible without the act of gifting back. It would have been impossible to think of the tool in isolation from its effects on the living land. The tool was conceived as *participatory*. The adjective captures the essential ontological status of the gift-giving tool. A gift-giving tool is conceived to facilitate a circle of participation, in full membership with both the human community and the more-than-human world. Here we see an example of a technological imagination that can truly call itself "ecological."

When a tool, such as a fish weir, possessed the inherent potential for overshoot, as well as for its equally corrosive power on the imagination, social practices would rein in the tool, buffering its effects on the land as well as on the imagination, weaving the tool back into participation. And what more effective way to do so than to impregnate the tool itself with gift-giving qualities—to gift the tool back to the river in its entirety, where it would become food again for others! And, of course, gifting back the actual physical structures would have been coupled with other practices of restraint, practices that absorbed more of the tool's corrosive potency, containing it further, and metabolizing it so thoroughly until what was threatening had been turned into a thing of beauty: a story to share, or a community celebration coming round and round again, year after year.

This leads to a question that is practically never considered in discussions about the feedlot industry's "sustainability": Is it possible to integrate the entirety of the feedlot industry's technological framework into patterns of cyclical participation with the more-than-human world? In other words, is it possible to integrate the entirety of the salmon feedlot industry—and, by extension, the entirety of the modern technological lifeworld—into a comprehensive gift economy? But is this question too abstract, too large? Are its implications too far-reaching? Then again: Why shouldn't we make a sincere attempt to answer it? Should we avoid it only because its political and social implications are potentially radical?

And what if the answer was, "No, it's not possible"? This would indicate that a feedlot industry (with its entanglement in fossil fuel extraction, its global consumption chains, its widespread use of plastics, its extraction of fodder fish from South American waters, or its dependence on agricultural products fertilized—again—with fossil products) cannot be sustainable. The imperative of reciprocal gift-giving makes claims along every part of the consumption chain, through every rippling wave of side effects, at all scales of place and time.

Let's remember that the aboriginal fish weirs were the most sophisticated of catch techniques available at the time. They were powerful enough to take all the river's salmon. This point cannot be overstated: Certain indigenous cultures had the power to overshoot their landbase entirely. The problem of drawdown technology did not originate with modernity. Like the closely related problem of collective maturity, it is older than modernity.

And again, drawdown technology is not a technological problem, but a problem of narrative. We are dealing here with an aspect of being human that might be unique among the multitudes of Earth creatures:[10] Like no other predator, humans have evolved infinite powers of manipulation. Humans can kill anything or anyone—other humans, other species, an entire river or mountain, the entire web of marine life, all forests, all land regions. Perhaps even the biosphere. The technological complex of the salmon feedlot industry only differs in degree from earlier, aboriginal drawdown technologies. Yet the feedlot industry has reached such a degree of complexity, leverage, and entanglement with the larger framework of modern megatechnology that it can be thought of as a travesty of the principle of reciprocal gift-giving. The salmon feedlot industry signifies a failure of the imagination, a failure of wisdom, a failure of story.

There is an opportunity in that failure—the opportunity to evolve rituals that could contain these infinite powers. It has been done before, and more than once! The indigenous fishing weirs of the Klamath and other Pacific rivers had degrees of that infinite power, as did the Ojibwa snowshoe, despite its ostensibly "primitive" nature. Within the reference frame of these technologies (a particular landscape, a particular species of prey), the people's powers were indeed infinite. They had the potential to kill their local cosmos entirely.

The difference between indigenous technologies and modern technology is not total; it is one of degree. Through the universalizing narrative of separation, and through modern megatechnology, the local cosmos now at stake is the living planet itself. It is within this finite local cosmos that our generation strives to absorb the forces of destruction unleashed by modern metaphysics, to contain them, and to metabolize them until they, too, have been turned into a thing of beauty.

Virtual Reality

The most parsimonious and precise way to think of the salmon industry would be as a recent chapter in a longer story of escape. It hinges on the notion of a disembodied mind in permanent exile, attempting to dominate all that it perceives as Other. The industry might not be fully ignorant of the larger ecological crisis convulsing through the biosphere, but its responses to that crisis come consistently from "within" narrative: It advocates more separation, not less; more control, not less; more manipulation, not less. It has no functional conception of the imperative of the return gift. Separation, control, and manipulation have been ideals since reason was first formally codified as a substance not entirely of this world. The salmon industry illustrates what Heidegger has called the "culmination of metaphysics": the Cartesian project appears to have reached a climax, a point of near-completion, in the modern technological framework. Within that framework and narrative, the world really does seem to increasingly resemble a *res extensa*. The ubiquity of modern metaphysics makes the story of escape seem natural, inevitable, and invincible.

But the evidence of systemic biospheric collapse persuasively shows that the narrative of escape is in crisis. The story of disembodied reason, governing an inanimate and voiceless *res extensa*, is becoming increasingly less plausible during a time when Earth agencies as potent as ocean acidification are stirring. Earth, long thought to be silent, has begun responding. Some responses are so composite, so brutal, so nonlinear, and point so very deeply into geological time, that they cannot be fully comprehended, nor predicted, nor contained, nor managed.

The industry's advocacy of more of the same—more separation, more control, more manipulation—might be another sign that the narrative of escape is in crisis. Reason might not be faring so well after four centuries of self-imposed exile. It might already have calcified and become gradually less able to respond to crisis creatively and resourcefully. Reason might already be in dire need of renewal and rejuvenation, and of being guided back into participation.

If reason is most at home inside the metamorphic depth of the biosphere; if the biosphere is now convulsing with collapse, as it seems to be; and if the storytelling animal would need to respond to the collapse with the greatest creativity and the greatest resourcefulness, then this spells real trouble. At the moment when it seems that reason needs to be at the height of its powers, it is instead rather badly stricken. With systemic changes under way, it would be foolish to write this off as a passing phase. The hardening of reason and ecocide are entangled in a cycle of positive feedback: The hardening of reason contributes to an impoverished sphere of life, which in turn offers fewer opportunities for reason to be brought back into a more reciprocal participation with the body and with the more-than-human world.

In all this, the salmon industry and its awe-inspiring success rates are nearly entirely virtual. Fueled by drawdown energy, its complex infrastructure is not integrated within organic cycles of participation or gift cycles. It is constructed in spite of them and in opposition to them. The industry cannot conceptually fathom the necessity to feed back to the land that which it has consumed. It has no working conception of an original agreement with salmon, nor with any other living creatures. It tears Being out of its richly textured, rhythmical self-birthing, which intrinsically comes with a need to withdraw, to retire, to decompose. The salmon feedlot complex generates waste— excess CO_2, plastics, dead zones. But waste is an ontological category that only exists in modern metaphysics. It has no place within a truly living biosphere. Inside a living Earth, there are no externalities. Each thing, each actant—whether it is excess oxygen, excess carbon, excess solar radiation, or any other disturbance—exerts adaptive pressure on all else, until it is brought back into loops of participation that are mutually beneficial.[11]

This not true of the disturbances or externalities caused by the feedlot industry. Structurally, they unbend cycles of participation. They cause loops to disintegrate. They rip lines through the composed thick of the biosphere. And they offer fewer opportunities for the embodied mind to be drawn into a state of awe and for reason to structure itself in accordance with specific places and seasonal rhythms.

For all these reasons, any nods to "sustainability" attempted by the industry must be considered only lip service. To craft technology that helps humans dwell here inside this Earth, inside this atmosphere, as participants in this hydrogen cycle and carbon cycle, we must ask how technology impacts our relationship with Being as such. "Technologies of inhabitation" would be those that have a participatory quality: tools conceived to facilitate gift-cycles throughout every step of their designated production, use, and disposal. If a community succeeds in embedding such gift-giving technology into its wider social practice, its members have every possibility of being attentive collaborators, rather than nuisances, in the biosphere.[12]

The salmon feedlot industry in its present articulation does not sufficiently problematize this charged relationship between technology and Being. It does not sufficiently question the basic metaphysical problem that the living Earth is being turned into an expanse of raw material. It is not grounded in the imperative of the return-gift. And it has no long-term solution to a problem it is not even aware of: namely that mind appears to be ever more starved as conceptual and physical boundaries continue to rigidify. The industry obstructs the work of creative adaptation necessary in this tumultuous time of transition.

It is not clear how the transition will play out. Even as modern infrastructure cuts deeper into seabeds in search of more pockets of stored fossil sunlight, even as new means of manipulation poke deeper into the genetic memory of fellow creatures to streamline evolution by the linear geometry of progress, even as engines unabatedly exhale more CO_2 into an atmosphere and into oceans thought of as dead dump sites, die-hard epistemological optimism is blindly calling up some of Earth's more frightening agencies, some of whom had been slumbering during the relatively stable interglacial period of the past ten thousand years. Tipping points are reached in the climatic balance—in the waters, on land—catalytic events that bring rapid and comprehensive shifts to what was thought to be a passive background, an inert *res extensa*. Within the very narrow narrative horizon of modernity, it is difficult to gauge the scale of the forces unleashed and the power of these agencies to act. But more insistence on control, on management, on certainty, on upholding the sharp separation between reason and the larger biosphere, will not lull these animate powers back to sleep. It will only infuriate them more.

What if, despite all this, the salmon are already laboring to bring the human animal back into a more reciprocal participation with Earth? What if they are already influencing the wounded thinking mind, offering to guide it out of its long quarantine? They might already be speaking to those two-legged apes on land—those of us who, like many others, are so drawn to their delicious flesh.

You seem a bit lost and helpless at finding your way to where you belong. Let us give you a fin. We know a thing or two about navigation, so we might be able to help you steer out of your loneliness. Relinquish that self-centered sense of control and entitlement. Invite that embodied mind of yours to embrace its own vulnerability. Let us feed you on terms our nations have agreed on together, not on terms imposed upon us by you. Recognize us in the uniqueness of our individual lives. Accept that there are times when we choose to feed you, and then there are times when we must hold back. Remember our obligation to feed not only you but many others, too. Remember your obligation to make return gifts to us. Give us the gift of your full-bodied attention, your curiosity, your feelings, your intuition. Gather your senses before you decide to take some of us, for then it will be easier for you to celebrate and regulate the links that connect our species. When you use tools to catch us, embed them into social practices that absorb any potential that you might become too proud. Integrate all your actions, and all your tools, fully into gift-giving cycles. Review the complex ways in which we salmon still enrich your embodied mind with metaphors, with concepts, with insight. Gift us with good stories and thoughtful practices, with careful speech, with song, with dance. Reclaim your sense of being an embodied intelligence, here alongside us, inside this living Earth.

CHAPTER THIRTEEN

Drawn Inside Geostory

The water you drink is three billion years old, give or take a few million years. The stuff your body is made of is at least 10 billion years old, probably older, and has been as far away as 100,000 light-years from where it is right now. The air you breathe has, in the course of its travels, been literally everywhere on the planet, and has slipped in and out of the lungs of almost every human being who has ever lived.

Would you act differently if you had a visceral sense of how eternal and infinite you are? What unprecedented behavior might you express? Visualize a waking dream in which you remember the water you floated in three billion years ago. Imagine you can see the light that shone on you 100,000 light-years ago.

ROB BREZSNY

*W*hat could it mean to transition out of the old story of separation and into a radically different kind of narrative? Neil Evernden has cautioned that "there is no way to deliberately elaborate a new story." One cannot "write a new story," but one can "listen to one." At best one can hope to "pull back and see what emerges to fill the void."[1] With each of the preceding chapters we have pulled back different layers of the story of separation. What is struggling to emerge in the old story's stead?

In the story of separation, a disembodied mind escapes into permanent exile, attempting to dominate all that it perceives as Other. The impulse for escape originated in a time of narrative ambiguity, when both tradition and the senses appeared to have been misleading the mind for a very long time. The distrust of the senses and tradition became codified and institutionalized. This can be considered the opening act of the modern story of separation, what Bruno Latour calls the "great narrative of emancipation."[2]

This present time is also one of narrative ambiguity. In attempting to eliminate all uncertainty, the story of escape provoked agencies that have caused an uncertainty of existential proportions. We can no longer take it for granted that the planet will survive the current shockwaves of anthropogenic disruption. Gaia's death is not impossible. The planet might have overcome all major disruptions in the past, but this is no guarantee that it will inevitably do so again. The planet is older now, the sun hotter.

For this reason, this moment of transition between stories is a bottleneck, a time at once fearful and laden with possibility. Arthur Miller, the playwright, once said that "an era can be said to end when its basic illusions are exhausted."[3] The modern ideal of mind's escape from the sensuous world is such an illusion, and the story of escape is becoming exhausted. Humans can thrive only if the larger Earth body does not suffer. The story of escape has undermined the very conditions of its own inherent optimism, leaving behind the kind of narrative void that Evernden foresaw. This final chapter addresses that void.

Latour has offered an intriguing name for a genuine alternative to the story of separation: "Geostory," as he calls it, would be "a form of narration inside which all the former props and passive agents have become active without, for that, being part of a giant plot written by some overseeing entity."[4] Latour mobilizes "earthbound, incarnate science" to the enormous task of making the human collective responsive to agencies acting on scales that all but shatter modernity's short-range attention span. Through such science, the collective imagination could grow so wide, so inclusive, so sensitive, and so deeply into time, that humans might learn to embed themselves inside Earth's larger storied unfolding. Given the forces already unleashed and the losses already suffered, a radically more inclusive narrative perspective offers the best alternative to the egocentric, hubristic, and short-sighted story of modernity.

I do not find it convincing that science alone is up to the task of integrating the human collective into Earth's ancient story. Next to science's piecemeal work of making us more keenly responsive to the systemic changes now under way in the biosphere, which is Latour's most immediate interest, there is also a complementary need: to pay close attention to our direct bodily experience, and to the way in which such experience places us squarely inside the metamorphic depth of the presence.

This seems like a mismatch: Here is the immense geostorical expansion of which Latour speaks; there, the kind of sustained attention for the presence implied in embodied experience. The mismatched scales of these two concerns complement one another: Each is able to draw the attention out of narrow and selfish concerns and into reciprocal participation with the land, as well as into processes that unfold at scales beyond our individual lives or imaginations. The work of narrating geostory is a participatory, animistic process between humans and Earth, a process of gradually letting ourselves be drawn further inward: into a felt kinship with the wider and wilder life of this planet, into time's incredible depth, and into the equally spacious depth of the presence.

"Geostory" is not restricted to the domain of science; it rather invokes a truly holistic perspective on knowledge, including science and philosophy, but also artistic expression, contemplative practices, on-the-ground action, and also a genuine respect for the insights of various indigenous wisdom traditions. Each becomes relevant in this larger generational effort of re-narrating the human more complexly and accurately inside this storied and ancient Earth.

The inherent thrust toward a rehabilitated sense of inwardness is what makes "geostory" a genuine antidote to the story of modernity. Through a geostorical lens, the four-centuries-old story of separation appears to be but a relatively fleeting intermezzo. The ancient, embodied experience of living inside a wombish world is gaining new credibility, endorsed by a renewed attention for the life of the body and for "earthbound, incarnate science." Both validate the primordial experience that we live, not on an inert piece of rock, but inside of something weirdly animate. Both give new substance to the old intuition that our individual lives are deeply entangled with the life of this larger sphere.

Under the Moon, inside Geostory

Widespread uncertainty and fearfulness are not the only parallels between this present time and the days of early modern thought. Latour has described our present historical moment as another "age of discovery," comparably radical to the developments of sixteenth- and seventeenth-century Europe.[5] Copernicus, Kepler, and Galileo helped explode the notion of space, leading to such philosophical responses as the postulation of extended stuff and thinking stuff, the world as a pure exterior separate from the inner life of the soul. It was during those days that the alleged view from nowhere gained credibility. It was also the time when it seemed possible to dwell inside a certain story from nowhere, to live by a narrative that was "not all there." That first age of discovery was closely allied with an impulse to externalize the material world, to disengage from it, and to stand apart from it.

In this present age of discovery, Latour suggests that we are seeing multiple developments toward a renewed experience of inwardness.[6] One which is particularly exciting is the explosion of the notion of time. Geostory unfolds on immense scales. Nowadays, we can think thoughts no earlier generation could think, such as the idea that all of civilized history has taken place within one brief interval between two vast glaciation events, each a one-hundred-thousand-year heartbeat within a longer pulse of glaciations, indicating that the living planet might be experiencing increased heat stress in the face of an older, more radiant sun. Or another: All of human history is but a footnote in the multivolume story of Earth's struggle to sustain life despite repeated catastrophes, such as Snowball Earths or the previous five major extinction events. Or the fact that the entire span of multicellular life on land is dwarfed by the vast stretch of time in which the oceans were the exclusive haunt of the microcosm. Or that Earth's moon appears to have been born at that terrible early hour when Theia, a rogue planet the size of Mars, crashed into the infant Earth, a sudden impact so apocalyptic it defies the imagination.

Such well-documented episodes help us re-acquaint ourselves with this planet in all its storied magnificence, ferocity, vulnerability, and resilience.

Latour speaks of this present explosion in our temporal horizon as a "geostorical extension,"[7] in clear parallel with the early modern spatial extension. Our perceptive horizon is being blown wide open, as it was then. The geostorical extension radically reconfigures the very parameters of what it could mean to be storytelling animals here inside this planet, at this vulnerable moment in the planet's life history.

———

Whereas the explosion of our spatial imagination was mirrored so clearly in philosophical responses to externalize the world, the explosion of our temporal horizon prepares us for the unique possibility *that we can be drawn deeply inward again*: into a deeper and more complex integration between humans and Earth, into a sense of interbeing that is experientially informed and scientifically sound. In this, Lovelock's Gaia theory might yet play a key role. For Gaia theory allows us to reclaim Earth as the inside that was forfeited at the dawn of modernity.

Latour observes a strange and rather beautiful reverse symmetry between Galileo and Lovelock. Lovelock's discovery of a self-composing biosphere is of such profound consequence that "very soon, in the history of science as well as in the popular imagination," it could stand side by side with the historical moment when Galileo overhauled the geocentric cosmic model.[8] After Galileo had aimed his telescope up into the night sky, Earth seemed *more* like all the other falling bodies than ever before. It seemed *less* unique than ever before. It became that notorious *res extensa*.

Lovelock did exactly the opposite. Already a famous inventor in the early 1960s, he was hired by NASA to help build a gadget that would detect life on Mars. Quickly, Lovelock began to suggest to his superiors that there was no need to send robots to the Martian surface. He suggested a simpler, more holistic solution. He reasoned that, in all likelihood, life on another planet could be detected without going there. After all, life on Earth had profoundly altered the atmosphere, using it as a source of raw materials and to exchange gases, making it a highly reactive atmosphere that was far from chemical equilibrium.

He also understood that the atmospheres of our two neighboring rock planets, both Mars and Venus, are both near equilibrium and thus highly unreactive. Stephan Harding, who collaborated closely with Lovelock, has described the atmospheres of Mars and Venus as "like a party where everyone is utterly exhausted and has gone to sleep, when no one has any energy left for conversation, for the exchange of addresses and phone numbers, for dancing, and for invitations to further parties."[9]

Lovelock eventually realized that Earth's atmosphere was unlike anything we know in the universe, precisely because it was so improbably reactive—"much like a party in full swing where people have plenty of energy for dancing and for lively conversations."[10] He also found documentation showing that the amount of oxygen in the atmosphere had remained roughly constant and at habitable levels for the truly enormous stretch of three hundred million years. But how could that be the case, he wondered, if free atmospheric oxygen and methane react with each other within days, producing carbon dioxide and water? This thought became the life spark for Gaia theory: Life on Earth seemed not only to have created the atmosphere, but to have regulated it over vast stretches of geological time. Life kept the planet habitable!

After Lovelock had moved his gaze from the sky back to Earth, Earth suddenly seemed far less like any other falling body. Our planet seemed suddenly far more unique than it had been. "From his tiny office in Pasadena, like someone slowly sliding the roof of a convertible car tightly shut, Lovelock brings his reader back to what should be taken, once again, as a sublunary world," Latour writes.[11] Once again there is something under the moon unlike anything else we currently know about in the cosmos. Latour actually calls Lovelock's discovery of Gaia the genesis for a "new form of geocentrism,"[12] driving home the reverse symmetry between Galileo and Lovelock. A view from nowhere is no longer adequate to recognize what it is like to be down here, under the moon, inside Earth. Because Earth is the only planet we know of that is "actively maintaining a difference between inside and outside,"[13] it is more accurate to say that it is like something to be this Earth. The biosphere possesses a certain inwardness, or subjectivity, that we have not yet observed anywhere else in the universe.

The scientific image of Gaia, says Latour, is arguably the one image that most sharply clarifies the dissimilarity between Earth and all other known celestial bodies. No other planet or moon that we so far know about can be said to have a life history, or biography, or geostory. None has evolved tightly integrated agencies the way Earth has, agencies whose concerted actions have given rise, across the span of geological time, to an increasingly complex living web able to maintain the environmental conditions necessary to thrive. No other celestial body, as far as we know today, has been able to actively keep its surface regions cool in the face of its aging (and increasingly hot) star, or to hold on to oceans' worth of water (even, as we saw, when the planet's gravity by itself is too weak to hold on to hydrogen atoms). In the expanding universe, Gaia—this "local ring of entangled feedbacks"[14]—is peerless. That is why he thinks of the image of Gaia as perhaps the most secular image ever evoked by Western science. No other appears to have recognized Earth more lucidly in its self-birthing, self-composing qualities.[15] No other takes us tangibly and concretely back into this biosphere, here under this moon!

Deep Time

Why is the universe not teeming with life, given that the ingredients for life seem to be so abundantly available in the universe? Imagine we are on Earth, sometime around 2,000 million years ago. Nothing here is like anything our mindful bodies would recognize today. There are continents, but they are fully devoid of life. There are no plants or animals or fungi anywhere, not even in the oceans. There is not yet a single multicellular creature on the planet. There is life, as there has been since life's origin some 1,500 million years earlier, but the first multicellular beings are still another thousand million years away. We are halfway between life's genesis and the origin of multicellular organisms. The bacterial ancestors of mitochondria might not have yet discovered the possibility of living snug and safe inside the single-celled bodies of larger hosts. The sun around which this younger Earth orbits is smaller and cooler. If you tried to breathe, you would immediately begin to cough, then suffocate. Within minutes, you would be dead. There is simply not enough oxygen in the air.

This is roughly the moment in the geostory when the global gift economy we call breathing first began. This chapter begins with a starvation crisis, which in turn leads to a planetary pollution crisis—the greatest pollution crisis ever, by far—which in turn leads to a reciprocal gift exchange of various gases that hold the atmosphere in a dynamic disequilibrium, to the benefit of all.

To sustain itself, life must constantly generate carbon-hydrogen compounds. Together with nitrogen, oxygen, phosphorus, and sulfur, these are the essential elements life needs to materialize out of chemical relationships. For 1,500 million years, ever since life's genesis, bacteria had been tapping into the abundant sources of carbon and hydrogen around them: They had drained the atmosphere of carbon dioxide, and they had learned to tap into every available hydrogen source, including hydrogen gas and the hydrogen sulfide that volcanoes were spewing out into the atmosphere. But now there was no longer sufficient hydrogen for everyone. The communities of photosynthetic bacteria across the planet were beginning to starve.

There was one source of hydrogen left, however, that no one had figured out how to use: dihydrogen oxide, also known as water—the stuff of the oceans. Water was super-abundant, but it was also an exceedingly difficult hydrogen source to tap into. Water's bonds between hydrogen and oxygen were so resilient that during these 1,500 million years, no microbe had yet learned how to break them. But need spurs creativity. Some cyanobacteria figured out the impossible—how to feed on water. "In an evolutionary innovation unprecedented, as far as we know, in the universe, the blue-green alchemists, using light as energy, had extracted hydrogen from one of the planet's riches resources, water itself," Margulis and Sagan write. "This single metabolic change in tiny bacteria had major implications for the future history of all life on Earth."[16]

But extracting hydrogen from the oceans produced prodigious amounts of leftover waste—oxygen. Oxygen was a potent toxin that threatened to kill all life. As it accumulated in the global atmosphere, many inhabitants of the young Earth died. This event has since become known as the Oxygen Holocaust. The word "holocaust" can be understood rather literally here. Derived from the Greek words ὅλος ("whole") and καυτός ("burnt"), the word means "the complete consumption by fire." "When exposed to oxygen and light, the tissues of these unadapted organisms are instantly destroyed by subtle explosions," explain Margulis and Sagan.[17] These early life-forms likely died due to internal combustion.

The frantic microbial community, under threat of being consumed by internal fires, began feverishly trying out, discarding, and exchanging defensive strategies. Eventually, there arose a few microbes who turned the problem into an opportunity: Rather than trying to defend themselves against internal combustion, they found a metabolic strategy that made the combustion a necessity. That strategy has since become known as breathing: "[Breathing] is essentially controlled combustion that breaks down organic molecules and yields carbon dioxide, water, and a great deal of energy into the bargain," write Margulis and Sagan.[18]

With the spontaneous emergence of the controlled combustion that is breathing, life closed a new cycle of participation that had not previously existed. Powered by the sun, some creatures now fed on water and excreted oxygen, while others turned excrements into gifts, using oxygen's fire in their favor, in turn excreting—gifting back—carbon dioxide and water. The Oxygen Holocaust inaugurated a whole cascade of creativity, leading to untold new food cycles, relationships, mutually beneficial dependencies. Out of a situation of peril and necessity, life spawned gift cycles of participation. This planetary gift exchange needed only the constant gift of solar light to keep everyone nourished. Life was no longer predicated on drawdown, but had learned how to integrate itself more deeply into the body of the Earth. In the process, it had also learned to participate indefinitely with the body of the sun. Life drank sun. It breathed sun. It ate sun. It embodied sun.

So, why is the universe not teeming with life, given that the ingredients for life seem to be so abundantly available in the universe? Aditya Chopra and Charles H. Lineweaver of the Australian National University recently made an exciting suggestion in an article published in the journal *Astrobiology* in 2016: "If life emerges on a planet, it only rarely evolves quickly enough to regulate greenhouse gases and albedo, thereby maintaining surface temperatures compatible with liquid water and habitability."[19] The fact that this rare event occurred on Earth is due to what they call the "Gaian Bottleneck." Here on Earth, life accomplished the far from obvious, far from certain feat of not going extinct! It's not the emergence of life that makes Earth unique; rather, it is the fact that life evolved so quickly and so creatively. What started out as life adapting passively to abiotic geochemical cycles transformed itself into life-mediated biogeochemical cycles, and this is what has made Earth a habitable planet for thousands of millions of years. This, too, Latour says, is what makes Lovelock's discovery of the scientific Gaia so remarkable. After Lovelock, there is now a solid scientific case that only Gaia has birthed itself, through an incredibly long succession of events that add up to one enduring, uninterrupted, four-and-a-half-billion-year-long story![20]

We still find ourselves "in narrative," except now we are finding ourselves thrown inside a narrative that did not originate with modernity, nor with early Greek thought, nor even with the distant origin of our own species. We are gradually learning to situate ourselves inside a narrative that unfolds across the huge expanses of deep time. This radically widened perspective, this geostory, might indeed offer an alternative to the egocentric, hubristic, and grossly short sighted story of modernity. In light of this, the collective craft of (scientific) geopoetics can be thought of, once again with Latour, as an assembly of "geostorians"[21] who endeavor to let themselves be drawn more accurately, and also more beautifully, into that much larger story. Now is a time when fresh forms of narration might begin to sprout from the wastelands left behind by the old narrative, forms of narration that help our tribe of storytelling animals to be drawn deeply inward again, into a sense of interbeing that is at the same time experientially informed and scientifically sound.

I had one of my favorite encounters with deep time on the Devonian coast in South West England, during the early summer of 2013. It was the first time I visited Schumacher College, and the first time that I joined ecologist-in-residence Stephan Harding on one of his Deep Time Walks. The walk was an in-scale excursion into geostory—each meter that we walked along the coastal trail corresponded to one million years in the history of Earth. At that scale, geostory stretched out over a 4.5-kilometer hike, taking us along narrow, steep coastal trails and ending at a charming tavern in a small fishing village. At times during the walk we all talked excitedly and impatiently. Other times we sat in solitude beneath rocky cliffs, letting the story sink in, or we broke out into spontaneous song while the rolling waves were adding their own beat. Or we looked skyward in awe and dread as the Late Heavy Bombardment was showering the young Earth with fireballs and frozen meteors for millions and millions of years. Somewhere along the trail that day, perhaps as late as the epoch when flowers and insects coevolved (the moment, Harding speculated, when the appreciation of beauty might have originated), it dawned on me: Walking the story beautifully activated all ways of knowing inside of me! My thinking mind was being challenged to ponder the weird and near-absurd events as they unfolded here around us. My body, meanwhile, seemed that much more capable of bringing perspective to the different chapters in the story. My muscles, skin, ears, and eyes all helped me find relevance, meaning, and value in the story. They helped me embody its unbelievable depth![22] I've made the same observation every time I've taken the walk since—in forests, along fjords, in a former landfill site turned nature reserve, and in downtown Buenos Aires. The mind on its own would be rather incapable of recognizing the story's qualitative aspects. It needs the body's gentle guidance to step inside, and to know why it matters.

At the end of the hike, Harding used a tape measure and placed the story of humans in relation to the larger Earth story: Wise apes as a species appeared somewhere between *20* and *30 centimeters* before the walk ended. The most recent glaciation period, after which the first domestication revolution began, ended *1 centimeter* before our hike was over. René Descartes dreamed his dreams *0.4 millimeter* before present-day.

I felt free to be impressed.

Turning Inward

If Gaia actively maintains a difference between inside and outside; if the biosphere possesses degrees of entanglement, inwardness, and subjectivity; if, further, Earth's degrees of entanglement, inwardness, and subjectivity appear to have been gradually *increasing* and *deepening* across the vast stretches of deep time; and if, finally, we can reasonably assume that the biosphere was never so profusely and wildly self-aware as it was at the advent of this most recent mass extinction—then it follows that we cannot truthfully know the biosphere through quantities alone. Entanglement, inwardness, and subjectivity are qualities, and we can know them precisely because we, too, are entangled within the biosphere, because we, too, have experiential knowledge of being subjects, and because our breathing bodies, too, actively make a difference between inside and outside. Like the salmon, we are insides within insides within insides.

This is how the sphere of life here on Earth has organized itself: Its interiority is complexly layered, corporeal, multicentered, porous, creative, self-birthing, autopoetic, and, importantly, it is always concretely localized. And so the "views from somewhere" conceivable now, in the wake of Gaia theory, are not only "views from inside this enveloping atmosphere." They are also innumerable variations on "here inside this specific bioregion," "here inside this animal or plant or fungi or microbe body," as well as, in our case, "here alongside these other-than-human sentient beings." Every subjectivity is a fractal, or incarnation, or incantation of the biosphere's wider interiority. The territory or terrain in which our sensitivity for the biosphere's qualitative aspects becomes molded is also the terrain into which we are situated with our bodies. From breathing forests to ripening seeds, from the swift hastiness of storm clouds to the seasonal return of winged migrants, the primordial instructors that tutor the human animal in Earth's stratified, complex, and storied interiority are places! And so the particularly human sense of self-reflexive interiority does not blossom at large within the larger Gaian interiority; it is nested: inside the atmosphere, inside bioregions, inside our bodies, alongside other sentient beings who possess their own sense of interiority. Each sensuous body and each place imbues its own flavoring to the experience of the real.

Where the Cartesian tradition took our experience of subjectivity as testimony to our existential alienation from the world, and as sanction for the sweet reverie of world dominion, we now know that this very experience of subjectivity actually embeds and interweaves us more deeply with this whirling sphere of life. As we rediscover the biosphere in its myriad qualitative aspects, we liberate ourselves from the futile attempt to capture the tragedy of ecocide in quantities alone—in cold statistics, cold graphs, cold language. To try and know this Earth more truthfully becomes tantamount to letting ourselves be drawn more deeply inside. It becomes tantamount to *turning inward*: To really sense, feel, and intuit into this planet's entanglements on the scale of our directly felt lifeworld. To touch the whirling interiority of this world with our sensitive skin, to smell it through our discerning noses, to see it through our acutely alert eyes. To "live in its presence and drink the vital heat of existence in the very heart of reality,"[23] as Catholic philosopher and mystic Pierre Teilhard de Chardin puts it. Turning inward: reaffirming this most concrete and experiential scale to reclaim something more than rationality. Turning inward: practicing an alertness to the real that can overcome at last the century-old exile of the embodied mind.

And as we are drawn more deeply inside, we finally come to know that it matters to Earth to go on living in diversity and abundance; that the dying around us is a manifest tragedy to Earth, that more diversity and more life are tangibly more desirable to Earth than more privation and more dying. We know that the idea to let a species go extinct at the expense of an extractive industry is demonstrably flawed, reckless, imprudent— that it is unambiguously evil—because the loss of wild salmon's unique subjectivity would be an irreparable impoverishment of the real.

One writer who can bring distinct clarity to what it might mean to turn inward is the anthropologist Richard Nelson. His memoir *The Island Within* traces his journeys on an unnamed island not far from his home in Alaska. In his book he probes at length the possibilities to craft an embodied epistemology, to speak a language of relationship and interbeing, striving to articulate clearly a growing sense of intimacy, where the boundaries of perception, and of language itself, become porous, permeable. Thus he writes:

> I become absorbed in the process of moving quietly, staring ahead
> through the variegated leaves and branches. I feel the air against

me, like a body of clear gel—an invisible flesh that fills the space inside the forest and covers the hard bones of rock underneath. The maze of tree trunks, branches, boughs, and needles penetrates the flesh of air as a web of veins. I move through them like a microorganism swimming inside a huge animal. I touch the spruce bough and sense it feeling me, as if it's become a nerve inside my own body, or inside the larger body that encompasses us. Just as the branches stimulate my senses, I stimulate the senses of the forest. We move within each other and feel each other's movements. During these moments, the notion of separating myself from the forest seems as untenable as crawling out of my own skin.[24]

His words seem to have been gathered attentively from within the various regions of his sentient body; they speak of an involvement with the presence so immediate and exposed that conceptual boundaries seem altogether to dissolve, opening up instead for a fluid participation with every felt detail. But in this deep entanglement, Nelson does not lose himself as much as he appears to be expanding. His words trace that expansion, but it is not so much an expansion outward as it is a growth into a more comprehensive and inclusive sense of interiority, or inter-being. The metaphors he chooses to trace his passage are metaphors of the body. In his raw exposure to the presence, everything spontaneously shows itself to him in its own life, and everything partakes of a larger life. He also writes: "The sense of *life* in this temperate jungle is as pervasive and palpable as its wetness. Even the air seems organic—rich and pungent like the moss itself. I breathe life into my lungs, feel life against my skin, move through a thick, primordial ooze of life, like a Paleozoic lungfish paddling up to gasp mouthfuls of air."[25]

His experience is one of being immersed within a *living immensity*, where his own self is not a separate "entity" as much as it is a cell or a microorganism, living "inside the larger body that encompasses us." His immersion in this larger body is fluid and dense, so much so that he finds himself moving "through a thick, primordial ooze of life"! And there is sentience, or subjectivity, inside that thick ooze: "Just as the branches stimulate my senses, I stimulate the senses of the forest." To sense, for

Nelson, is also to *be sensed*: "We move within each other and feel each other's movements." Just as he cannot deny the reality of his own sentience, it is impossible for him to deny the sentience of the encompassing forest.

Not unlike Abram's writing, Nelson's sensual speech is not derived from a series of deductive reasoning, nor could it be neatly abstracted into concepts. Like Abram's writing, Nelson's writing bears witness to the fluid and reciprocal process of perception itself, which is the activity that constantly negotiates between our private sense of interiority and the larger interiorities that surround us. Nelson evokes a sense of embodied continuity that slips effortlessly across the various layers or strata of interiority, from his own body to the surrounding forest. In some sense, we can say that we are witnessing, in Nelson, how his keen alertness to the process of perception itself births a fluid and utterly porous style of thinking! We are witnessing, in a way, an ongoing process of discovery and insight that flows from the land and through Nelson's body, leaving breath tracks on the page.[26] The printed words then become signposts; they point toward the possibility that such fluid, participatory alertness to the real is in principle possible at any time. They point toward an observation that Abram made in that eminent final footnote of *The Spell of the Sensuous*, namely that reflection remains "[rooted] in such bodily, participatory modes of experience."[27]

The plain beauty of Nelson's writing should not distract from the fact that it also launches a concrete critique against the particularly modern habit of separating mind from body. This is not an overt critique; it is more straightforwardly an account of his lived immersion, and so his rejection of the Cartesian split is not directly manifest. And yet it undoubtedly forms the historic-philosophic backdrop of this book-long attempt to carefully sound out possibilities for more integrated ways of thinking, and of speaking. Thus he is able to say, closer to the end of the book: "My purpose, which has emerged gradually and of its own accord, is to understand myself in relation to a natural community of which I am, in some undefinable way, a part . . . not only as a visitor, but as a participant."[28] At that point he can also write, remarkably, that "the exploration has turned inward . . . The island is not just a pleasure to my senses—it is my home, my ecological niche, my life broadly defined."[29]

Turning inward: Here is the theme where the kind of phenomenological thought practiced by Nelson and by Abram converges directly with the work of Latour. All of them work to help create a moral imaginative space where the human can be experienced again in richer and more reciprocal participation with the larger living world. And this, I suggest, is the central motif of geostory.

Nelson's writing also illustrates how such participatory alertness sculpts the experience of time. Ever crisscrossing the island's windswept rain forests, Nelson finds ample opportunity to ponder the lives of the trees among whom he walks. And as he does, time itself seems to acquire a certain layered viscosity, a textured thickness that extends fluidly from the presence and into the sediments of the past, as well as into possibilities for the future to shape itself. He finds himself drawn into a much-expanded sense of presence, pondering the fate of the "trunks and boughs and branches that have fallen onto this earth for thousands of years," and "the little showers of needles that have shaken down with every gust of wind for millennia."[30] His mind is being beckoned into this larger awareness through his body's sinuous participation with the forest. Nelson finds himself in "this forest of elders, this forest of eyes,"[31] surrounded by a "gathering of ancients,"[32] himself a mere "upstart" who awaits "the quiet counsel of venerable trees."

Nelson's sensitivity for time's entangled, nonlinear depth clearly echoes Latour. Nelson is well aware that this island has been substantially wounded by recent clear-cuts, which turned larger swaths of the island into "an enormous graveyard, covered with weathered markers made from the remains of its own dead."[33] At one point he stands atop one of those "weathered markers" when he begins to count the annual rings of the massive stump beneath his feet, finding, after a while, that "the tree died in its four hundred and twenty-third year," cut down by a person who "would have seemed no more significant than a puff of air on a summer afternoon."[34] His encounter with the clear-cut stirs a near-insufferable tension. Here, on the one hand, is the ludicrous banality of the killing—perhaps the logger who took that tree, Nelson speculates, "thought only about the job at hand, or his aching back, or how long it was until lunch."[35] But, on the other hand, there is also the fact that those ancients who remain standing seem to have an uncanny power to share with humans "in a common nurturing."[36] The common nurturing of which he speaks is the cultivation of a sense of reciprocal obligations: "I am cautious and self-protective here, as anywhere, yet I believe that a covenant of mutual regard and responsibility binds me together with the forest."[37] As he finds himself being drawn into this larger and more complex experience of time, this is what takes shape: a duty both to honor, and to concretely defend, whatever parts of the forest are still standing.

Nelson's writing articulates a much-expanded sense of sensitivity and responsibility. He illustrates the kind of place-based concreteness that must be as much part of narrating geostory as locating the human adventure inside deep time. The radical temporal expansion (and entanglement) of which Latour speaks is fully complementary with Nelson's embodied, intimate attention for the here and now. Both are aspects of the same relational, expansive, and entangled alertness to the real that lies at the heart of geostory.

And finally, this: Even a landscape as severely wounded as the clear-cuts that Nelson walks through must not necessarily be dead forever. Nelson writes:

> I try to take encouragement from the ten-foot hemlock and spruce sapling scattered across the hillside. Interestingly, no tender young have taken root atop the flat stumps and mossless trunks. Some of the fast-growing alders are twenty feet tall, but in winter they add to the feeling of barrenness and death. The whole landscape is like a cooling corpse, with new life struggling up between its fingers. If I live a long time, I might see this hillside covered with the beginnings of a new forest . . . The whole community of dispossessed animals would return: red squirrel, marten, great horned owl, hairy woodpecker, golden-crowned kinglet, pine siskin, blue grouse, and the seed-shedding crossbills. In streams cleared of sediments by moss-filtered runoff, swarms of salmon would spawn once more, hunted by brown bears who emerged from the cool woods.[38]

So, yes. At least some voices can endure even in the midst of a wasteland. Even a graveyard can speak to the careful observer of what could be once again.

Two Dams Down

Fog can hover thick in early autumn, here where the river spills into the ocean. The Elwha once housed ten anadromous runs of native salmon and trout, among them steelhead, chum, coho, pink, and sockeye. It was also home to a mighty run of Chinook salmon, the King salmon. But when Elwha Dam was begun in 1910 and the Glines Canyon Dam was built in 1927, all but the lower 8 kilometers of the Elwha were blocked to the returning fish, denying them access to a watershed that includes a main river channel of 70 kilometers as well as 150 kilometers of tributaries. The people here remember those days when the dams went in. Before the dams were built, they say, the river would turn an intense red because of all the salmon who came up to spawn. After the constructions were finished in 1934, you would see a different red just downriver of the dam—the blood of those salmon who tried to return, unprepared for the possibility that their passage might one day be obstructed entirely. They jumped, and they jumped, and they could not overcome the concrete wall. They crushed their heads in numbers large enough to dye the river red.

They remember that once upon a time, the Elwha River produced enough fish to sustain their community year-round. Before the dams went up, the Elwha produced some four hundred thousand wild adult salmon every year. In the one hundred years that followed, those numbers plummeted to less than 1 percent of the original numbers. The watershed community, of course, was affected. "The exile of the salmon has meant the banishment of other animals that otherwise would feast on the fish. The area's populations of bobcats, bear, mink and river otter have likely declined," writes Abigail Tucker of *Smithsonian*. "Since salmon carcasses aren't fertilizing riverside vegetation with nutrients brought upstream from the ocean, even the cedars starve."[39]

Now it is August 2011, and a thick fog envelops the green banks of the river, spraying an indistinct white glow across the water. Pat John of the Ahousaht First Nation and Lower Elwha Klallam tribal member Mark Charles stand on the riverbank, raising their voices in song to the rhythmic beating of a drum. Next to them stands Charles's four-year-old grandson, Roger Tinoco Wheeler, and as the two men sing, the child tunes in to the song. He has listened to his grandfather sing to him for as long back as he can remember. Others, young people and elders, stand by to watch and to listen. Off on the side, Lower Elwha Klallam tribal member Rachel Hagaman and her sister Lola Moses are busy weaving rafts of cedar bough. The previous year brought home fewer Chinook than ever before, and this year only five Chinook were caught for the community's First Salmon Ceremony. These five fish are cut into pieces and distributed among the seventy-or-so elders of the tribe. Their carcasses will later be bedded onto the cedar rafts and set out back onto the river. This is how the Chinook are able to return to the river, back to their own people.[40]

Many have gathered here on the Elwha's gravel banks, tribe members and visitors, and the fog still hangs heavy above them all—trees, salmon, humans. Off in the distance the fog swallows the dark gray water, and the ghostly trees and shrubs, and the colors of autumn. Rachel Hagaman and Lola Moses have laid several of the Chinook carcasses on the cedar raft. Together with elder Ben Charles they carry the remains of the fish down to the water. The raft is seized by the current and starts bobbing off downstream. Charles offers some word of reflection: "I'm glad for all the reviving, the first salmon . . . our community welcomes the first salmon back, and sends that spirit of the first salmon catch where it needs to go. Tell them: 'These people love us, they need us, this is where we need to go, c'mon.' Then all kinds of salmon will come up this river."[41]

Chairwoman Frances Charles also seizes the opportunity to offer a few thoughts:

We are humbled and honored to be able to have the environment come back, we need to have the wildlife come back, the eagles, the beavers, eating off the salmon, you don't see too much of that anymore, and we are going to be able to revive some of that. . . . It's about our ancestors and those before them, we are walking in their footpaths, all the ones before us, they are the ones we want to recognize, and whose footsteps we are following. These were the foods our people lived off. The fish people were beaten and arrested for, to provide food for the table. How do I feel? I don't even know, we have been talking and dreaming about this for so long. . . . We were always told it would never happen. It is going to be an overwhelming day. I think about all the work, the effort, over all the generations. It's a process of restoring what was lost.[42]

Those who stand by to watch and to listen know that this ceremony will be the last of its kind. Under the circumstances, this is good news. The Elwha Dam and the Glines Canyon Dam are being dismantled. Starting in mid-September, the river will be given back to itself. It will be the largest dam removal ever, anywhere, up to this point. And to give the salmon a chance to reclaim the place they have been denied for a hundred years, the Lower Elwha Klallam have agreed to stop fishing for the next five years.

In the meantime, much work remains to be done. The Olympic National Park nursery is fostering hundreds of thousands of plant seedlings, whom they will plant to help recolonize the barren banks of the valley once the water runs freely. Timing is critical. With the soil exposed and vulnerable, invasive species might easily take over the uncovered ground. But if all is done well, the forest will be aided to reclaim land it once helped shape.

The human community, too, prepares for the transformation that is under way. No one alive today has seen what the river was like before the dams went in. The Olympic Park Institute organizes field trips, excursions, and camps, all aimed toward children, with the explicit goal of giving them all a chance to (re)connect. Old ways and new ways go side by side here. "We want them to think, 'Maybe science is something I could do,'" adds professor Robert Young, who played a role in securing the necessary funding for the dam removals. "We want them to say, 'I could be fixing this river. I could be helping it heal. I could be uncovering sacred sites. That can be me. And it should be me.'"[43] For one week, the middle-school children learn about life on the river. They learn about the science, they are guided on a vision quest, they learn about herbal medicine and native food plants. They also play by the river. Toward the end of the week, they go on an overnight canoe trip across nearby Lake Crescent, a hauntingly beautiful mountain lake some miles west of the lower Elwha. Dinner that night is a rich array of native foods, supplemented, as Abigail Tucker recalls, by teriyaki chicken brought over from the school dining hall. When at last the counselors say it is time to serve the finishing touch—salmon—there isn't any salmon there. Abigail Tucker recalls:

> [T]he counselors explained that they'd gone to the grocery store, where a single filet of white king salmon cost $60, and the program couldn't afford it. Instead, they fashioned a cardboard cutout of a chinook. Using the model, they explained how the Klallam might have smoked salmon in strips or boiled it in a bentwood box, eating even the eyes and cheeks. They demonstrated how the Indians would push a butterflied fish onto a split stick leaned over the fire, catching the ocean-scented juices in an oyster shell to drink afterward. The kids watched with wide eyes. Breathing in the wood smoke, one could almost taste pink, flaking meat.
>
> That night, the children practiced the welcome speeches they'd recite at the beach in front of their parents the next day, and the journey and greeting songs they'd been learning all week, which tribal members—grieved that the originals were lost—composed in the late 1980s and early 1990s for get-togethers with other tribes, and which typically have a strong rhythm meant to be banged out by drums or canoe paddles.
>
> They also sang one of the sole surviving Klallam songs, antique recordings of which date from the 1920s. All modern Elwha ceremonies end with its singing.
>
> But this is not a thumping, enthusiastic paddling anthem. The haunting "Klallam Love Song" is about absence, longing and the possibility of return. Young women sang it when their husbands were away. The words are very simple, repeated over and over. "Sweetheart, sweetheart," the women cried. "You are so very far away; my heart aches for you."[44]

As I write, the Elwha is running undisturbed again from its many scattered sources throughout the Olympic Mountains, all the way down to its estuary. The Klallam creation site, the place of coiled baskets, has emerged again from the waters, and the Klallam have returned to it. Native forests have begun recolonizing the two former lake beds. Salmon are spawning again several miles above the former Elwha Dam site. Already "birds, bugs and mammals are feasting on salmon eggs and carcasses as fish once again nourish the watershed."[45] Bears, cougars, minks, and bobcats have quickly learned to scavenge again on salmon carcasses, which for such a long time had haunted the watershed with their absence.[46] With the Elwha Dam and Glines Canyon Dam gone, 70 miles of habitat has been reclaimed for the steelhead and the five salmon species. In time, as the river rejuvenates itself, four hundred thousand salmon and steelhead might return to the Elwha every year. Christopher Tonra, scientist at the Smithsonian Migratory Bird Center in Washington, DC, speaks of the Elwha dam removals in awe: "It goes against my deepest notions of how fast ecosystem recovery can possibly happen. . . . We are all trained, as biologists, to think of things over the long run. I am not saying the Elwha is fully recovered. But it is so mind-blowing to me, the numbers of fish, and seeing the birds respond immediately to the salmon being there. It makes the hairs on the back of my neck stand up."[47]

Though not a member of the tribe, the *Seattle Times* journalist and author Lynda Mapes has long been a sensitive chronicler of the events unfolding around the Elwha, and in the Klallam community at large. As time went by, she also seemed to have noticed the way the river was beginning to "lay its claim upon her," as Abram formulated it. One can see it in her writing. Her words, like those of many others before her, have gradually become saturated with river life, and with the life of the senses. In her book *Elwha: A River Reborn*, she writes:

Water tumbled over a branch, driving silver bubbles of fresh oxygen into a plunge pool of cold, clean water. This was not the Elwha of the whitewater kayaking films, blasting furious through rock canyons. Here, drama played out on a small scale, as tiny sticks and bark caught in jams of branches that made eddies in the flow.

A juvenile fish no longer than the joint of my finger wiggled into the pool beneath the branch, facing upstream as a cluster of alder cones swished by overhead on the water's surface. Threads of algae streamed like hair in the current as the water eased downstream, alive with drifting seeds, detritus, and bugs. Leaf litter cartwheeled down the fine, sandy streambed. The shredded bits tumbled down a mosaic of pebbles on the bottom, overlain with the rippling reflections of sunlight. I realized as I watched that the woven pattern of the current, visible in the surface of the water, was mirrored in the stem and vein pattern of the alder leaves, the stem and branch pattern of the sword ferns, and the tiny branched shapes of the moss on the banks. Land and water interwove and interacted at every scale and dimension of the Elwha. . . . In the return of wholeness to the landscape, tribal members say, is the return of wholeness to the people. Not only for the tribe, but also for the rest of us.[48]

The act of restoring the Elwha to health is a concrete political act of healing a landscape, but it is also an important symbolic gesture, precisely an act of *restorying*. The dismantling of the two dams has initiated a reexamination of the various people's complicated relationship with the wild, and salmon are once more central to this work of restorying. Salmon are beings of flesh, blood, sentience, and intention, but they are also symbolic creatures, and humans create narratives from symbols. Dismantling the dams so that salmon can once again dash up the Elwha's glacial currents symbolizes hope and defiance, as well as love for the strange and exuberant Otherness of salmon. And it symbolizes a striving to create a more complex, reciprocal, integrated, and beautiful relationship between earthlings and their particular Earth region. It is part of the larger effort to dismantle the old story of separation.

The Elwha River appears to be offering its help: Mapes's words bear witness to the way in which the Elwha strives to make itself heard by humans, even after a century-long siege. Her words also indicate that at long last, the human collective is learning again to listen. And to respond.

The work of re-allying mind with body, as well as the work of mobilizing earthbound, incarnate, embodied science for the task of restorying the human-Earth relationship are complementary efforts. Each can help us widen our sense of relatedness with and responsibility for Earth. As such, they are two important entry points for the ongoing effort to dismantle the modern story of separation.

But the story of the Elwha points to yet a third—material, on-the-ground resistance to the old story. The theoretical and moral dismantling of obstructive forces needs to be matched by efforts to also dismantle such forces physically. Restorying the land needs to go hand in hand with restoring it, and vice versa. While some losses might be irreversible, others are not. Some of the harm inflicted can be undone. Some graveyards of old still speak to the careful observer of what could be once again.

The old story, which was "not all there," did not have the last word in the Elwha watershed. The Elwha resistance movement has succeeded in completing the largest dam removal project anywhere, ever. The explicit motivation behind the ambitious project has been to help the salmon rebound from near-extinction. The example of the Elwha illustrates the way in which the narrative adaptation out of the shortsighted and egocentric story of separation is already on its way.

But even as the Elwha watershed community is doing its part to heal from old wounds, larger forces are on the move, foreshadowing conflict to come. Recent years have seen a worldwide proliferation of large hydropower dams, at such a rate that one new, large dam has been completed, on average, every day![49] Norwegian companies are also expanding their production of Atlantic salmon into the Pacific Northwest, currently clashing with Ahousaht, Kelsemaht, and Tla-o-qui-aht First Nations, among others.[50]

Meanwhile, systemic changes are under way in the biosphere, and they do not leave the Pacific Northwest undisturbed. The year 2015 scratched the all-time record for lowest snowpack levels throughout the Pacific Northwest, causing the state of Washington to declare a drought emergency.[51] Atypically low snowpack levels, combined with atypically high temperatures, resulted in higher water temperatures in rivers throughout the region. This became a serious obstacle for salmon, who thrive in cold waters. The mighty Columbia River suffered salmon losses in the hundreds of thousands that summer, numbers that the Oregon Department of Fish and Wildlife described as "unprecedented."[52] Teresa Scott, Washington Department of Fish and Wildlife's drought manager, commented on the situation: "We are mounting short-term responses this year, and anticipating a recovery from these conditions in the near term, but certainly this is a wake-up call and a dress rehearsal for what fishery managers years from now will be dealing with on a regular basis."[53]

Recent evidence published in *Nature* also indicates that CO_2-induced aquatic acidification appears to be affecting not only oceans but also freshwater habitats. A team around Michelle Ou from the University of Vancouver exposed the larvae of pink salmon to acidity levels expected for the year 2100. Ou and her colleagues found a number of effects. Not only were the fish smaller, lighter, and less responsive to alarm cues, they also showed signs that their senses were impaired. They were less able to smell the signature amino acids that would help them recognize their unique birth-stream. The findings are so recent that it is difficult to draw any conclusions from them. Still, Ou and colleagues suggest that "future populations of pink salmon may be at risk without mitigation."[54] And furthermore: "Given this species' central ecosystem role in freshwater, marine and terrestrial food webs and their important economic and cultural roles in Aboriginal communities, continued increases in CO_2 may have widespread implications on ecosystem productivity and the many communities they support."[55]

Much originality, joy, resourcefulness, political cunning, and hard work is already being channeled toward returning place-based, inter-ethnic communities to wholeness. Places such as the Elwha watershed are now becoming pockets of healing and of resistance. New narrative strands are being crafted, strands that interweave place-based customs and rituals with contemporary science, as well as with the knowledge of the body, creating niches of immense inventiveness and fertility, as well as defiance against the story of escape.

Yet, systemic ecocide will undoubtedly continue to intensify in the future, at least for some time, as will the deep grief that is the signature of trying to live thoughtfully and compassionately in this time of loss. Our present generation is engaged, according to Lester Brown, in a "race between tipping points and political systems."[56] Fritjof Capra and Pier Luisi offer us a sober yet confident outlook of what might lie ahead: "Gradual changes will not be enough to turn the tide; we also need some major breakthroughs. The task seems overwhelming but not impossible. From our new understanding of complex biological and social systems we have learned that meaningful disturbances can trigger multiple feedback processes that may rapidly lead to the emergence of a new order."[57]

No matter how hurtful the present moment might be, this present remains a cradle of possibility. It is fully possible to hasten out of the narrative of escape and into the richly textured landscape of geostory, even as monsters are waking. If the extinction of so many fellow creatures depletes our own emotional and mental resourcefulness, then the opposite would be true as well: Actions that help restore the land, or community efforts that create opportunities for life to reassert itself, will not only heal the ecology of the land, they will also return greater resilience and flexibility to the collective human imagination.

Elders

In the story of Salmon Boy, there is that crucial sentence: *The boy started drowning but the salmon came to his rescue.* Psychologically, this moment depicts the perilous passage from adolescence into maturity. As an adolescent, the boy's imaginative horizon is too narrow for him to be able to understand the original agreement between his people and the salmon. He does not grasp the importance of participating in the gift cycle, the cycle of mutual obligations. Failing to do so, he suffers from a very real peril. His drowning is not only a drowning of the body but also a drowning of the spirit.

This boy represents aspects of that same smallness-in-largeness that the project of modernity has taken to the extreme: His self-absorption is coupled with his disregard for, and ignorance of, the long-term consequences of his actions.

But just when his egocentric actions threaten to kill him, the salmon assume the role of elders. Thanks to them, he can see the suffering he has caused. They give the boy the chance to hasten the death of the ego, and to be reborn, recomposed into a comprehensive sense of self. This new self is not atomized but relational.[58] It is coextensive with his larger human community and with the living land into which humans are bedded. In the story, the boy blossoms into maturity only because the salmon come to his aid, even after he has treated them disrespectfully. Through them, he is offered redemption. His psychological passage into maturity also brings practical benefits: When he returns, the boy enters the ranks of those who guard the original agreement through words and through deeds. He allies himself with those human elders who labor to keep the cycle of greater returns moving.

It is possible to think of the dying now under way as the rite of passage of our time. Planetary ecocide is a moment of narrative transition, and as such it remains also an opportunity ready to be grasped. The story of Salmon Boy shows that salmon have assumed the role of elders before. Could the salmon assume that role again?

In writing this book, it has felt that way. I dare say that it was the salmon's guidance that led me to certain punctures in the fabric of humanity-as-separation, and that helped me catch glimpses of what Latour calls "Gaia's narrative complexity and entanglement"[59] beyond the old story's frayed fabric. It was the salmon who intimated a number of pathways for encountering the problem of narrative appropriately: Follow loops of entanglement even across received disciplinary boundaries; practice forbearance against the temptation to reduce others to human concepts; practice nonhierarchical and relational speech; strive to unlearn habits of thought that hold humans apart from this vast, wind-swept Earth community; treat the text not as an assemblage of atomistic units gathered in a line but as an emerging, integrated, and necessarily provisional network.

My role has been as much that of writer as of listener; it has been that of an active participant, rather than disinterested observer. Salmon have not been an "object of study" in these pages; rather, they have conspired in the writing of it. Through this participatory exchange, I hope to have contributed a number of worthwhile reflections to the ongoing work of restorying and restoring the human-salmon relationship. I hope to have lent my voice to the larger cultural work of hastening the transition between stories that is now upon us all.

———————

Throughout this phenomenology of story, we have moved through the entangled confines of the narrative of humanity-as-separation and toward a growing recognition of other-than-human agencies that populate this same Earth alongside us. We are being drawn back into the living Earth by agencies other than human. This recognition is a key accomplishment in the context of this work. I have found support for Abram's contention that animism, or the notion that everything speaks, is a "wider and more inclusive term"[60] than rationality, coming before it, grounding it palpably in the soil, and the air, and the seas.

Much of this work has been about recognizing tensions in the encounter between the story of separation and its physical manifestations on the one hand, and contemporary insights into the complexities of the living Earth on the other. These tensions indicate that the old (so-called "modern") story is not true in a wider, ecological sense. Its nods to scientific authority are ideologically motivated. The kind of science that is being advocated from "within narrative" conveniently serves the project of power, but it is, like the story itself, provincial and narrow-minded. It is not to be confused with what Latour has designated as "earthbound, incarnate science." Earthbound and embodied science in the various disciplines is disclosing a world less mechanistic, less reducible to abstraction, richer, and more resonant with the indigenous knowledge of place-based cultures. We are once again finding ourselves situated deeply inside a living planet.

The provincial story of human exceptionalism does not withstand sustained critical attention indefinitely. Though the story's conceptual entanglement is dense, though its physical infrastructure is ubiquitous, though its ambition is indeed totalitarian, the story has been pushing hard toward a saturation point. Dissatisfaction is mounting, and the story's inherent inconsistencies are already causing it to implode. Change is inevitable. And that change might be both rapid and profound. There is a very different story that we are being asked to learn, and quickly. There is another story that strives to tell itself more fully through us. The narrative adaptation out of the shortsighted and egocentric story of separation is already upon us. Part of the work entrusted to this generation must be to hasten this changeover.

Like Descartes, Galileo, and other thoughtful chroniclers of the late-medieval collective psyche, we try to comprehend the scope of all that is unraveling before our eyes, now that the Holocene stable state appears to be fading. Like them, we struggle to respond creatively, and meaningfully, and also beautifully, as received certainties are failing to give guidance. But unlike them, we choose to emphatically resist the ever-looming temptation to exert more control, more separation, more abstraction, more domination. One need not condemn their failure in anticipating the nightmares that would come back to haunt the heirs of modernity. It is sufficient to recognize that some of their responses to crisis were mistaken. It is our turn to try and make better choices this time around.

As in the story of Salmon Boy, this is the moment in which the culture of the ego must die. If we were to succeed in passing through this critical transition—and still have a living planet—we could be reborn into a far more comprehensive sense of being human. We could be mature Earth citizens, ready to assume the responsibility that comes with being such flexible, undetermined, endlessly creative, and wise apes.

The work of unweaving the story of human exceptionalism has begun, and if this work has succeeded in causing some meaningful disturbances, I am content (for the time being). We see through the old story's worn fabric, and we find that we are deeply embedded within a far larger story. That larger story is awe-inspiring, terrific, and very beautiful. But it is also able to frighten and destroy. This book does not lend itself to any notions of romanticism or escapism. There is no invulnerable place to escape to any longer. The only refuge is the living world itself, in its fragility, its unruliness, its unpredictability, but also its beauty and wonder. This is a work of resistance, and of homecoming, and of asserting life's irreducible fullness.

And how spacious geostory is! What pressure it can exert on received frames of reference! The salmon industry, now so dominant an actor in mediating human-salmon relations, has been around for about half a century. But humans have been asking to receive the gift of the salmon's flesh at least since the last glacial period ended some ten thousand years ago, when the oceanic rims of northern Europe and America became more accessible for human settlement, as well as for the salmon. As the glaciers retreated, it is likely that the newly accessible northern regions both of Europe and America saw humans and salmon arriving side by side, coevolving strategies to dwell inside emerging habitats. What is half a century, compared to ten thousand years? One part in two hundred!

Salmon, as the distinct species we know today, have been accepting the responsibility to return from the Pacific and Atlantic Oceans for at least six million years. Six million years equals thirty times the life history of Homo sapiens as a species. It is also three thousand times longer than the time that has passed between today and the life of Jesus. And it corresponds to fifteen thousand times the time that has passed since some thinking apes first thought it wise to treat Earth, and all her processes, and all her creatures, as a machine. Given the salmon's very long sojourn inside this ancient planet, it would seem to be far more parsimonious and precise to think of salmon not as objects, not as commodities, but as elders.

The Story of the Smolts

I depart as air, I shake my white locks at the runaway sun,
I effuse my flesh in eddies, and drift it in lacy jags.
I bequeath myself to the dirt to grow from the grass I love,
If you want me again look for me under your boot-soles.

WALT WHITMAN

There were once three adventurous smolts named Copernilax, Galilax, and Desolax. All their lives they had lived inside their birth river. The river was their local cosmos, their wombish world. It was finite, it was knowable. Their bodies recognized its every sound, its every gurgling and splashing and gushing and sloshing. Ever since the beginning of their young lives, they had learned to trust the river, and they had learned to trust their senses. They knew the song of the river. It was as it was.

There came a night when they knew. They deserted their river bottom territories and began to drift. They abandoned themselves entirely to the river's guidance, breathing it, letting its drift become the measure of their imagination. They could not have imagined what lay ahead of them. How could they have anticipated that beyond their river, there was an entire *ocean*? There was no way they could imagine the huge expanses of water beyond their river cosmos, that vastness that just stretched on and on and on, so far that neither their eyes, their ears, their tongues, nor their sensuous sidelines would be able to penetrate it.

Changes were adrift inside the watershed. Copernilax was the first who felt an urge to herd the others into a tight school, even though most of them were clearly not yet interested. It was Copernilax whose skin was first to turn silver, when all the rest of them were still river-shaded. His body became larger and slenderer and stood out awkwardly among the rest, which was the last thing any of them wanted at a time when the school of duplicate bodies was their only refuge. In fact, Copernilax had been visited by strange dreams of late. Dreams of huge flare-ups, of a river cosmos that swelled and swelled until it all erupted in a massive SPLASH! He had tried to tell the others about his dreams, but they had said, "Look around, you fool! Here's a riverbank, there's another, and between this and that is the cosmos!"

One early evening, a grey heron stood in the shallow stream. Swarms of mosquitoes danced just inches above the water. Swallows were whizzing to and fro. Heron saw a tiny shape pass by, just below the black surface. His beak darted forward, quick as a flash of insight, and he swallowed Copernilax whole.

The next morning, Galilax woke up before everyone else. A strange sensation inside his mouth had woken him. Had it been a dream? He remembered feeling curious and anxious. It felt as if a faint memory had dawned inside of him. Suddenly, he was overcome by haste. He couldn't wait for the others to wake up. A tremor shook his body, and he dashed downstream. There was no time to lose. Soon he disappeared beyond the next bend in the river.

Meanwhile, Desolax awoke. He'd been dipping in and out of sleep himself that night, pondering, in his own dreams, a strange question. *Quod vitae sectabor iter?* (Yes, Desolax dreamt in Latin.) What path in life shall I follow? "Indeed," he thought, "that is the question." And then he said to himself, for the others were still asleep: "It does rather seem like an appropriate question, here on this journey toward an uncertain destination." Just then Desolax saw Galilax disappear beyond the bend in the river. What was Galilax up to? Desolax dashed after him. He rounded the bend in the river and—almost bumped into Galilax!

There Galilax stood. His muscles were shuddering. His fins, his scales, the whizzing network of his nerves—his entire pulsating body seemed strangely on edge.

"Copernilax was right all along," Galilax said at last, not turning to look at Desolax.

Just then, Desolax was knocked over by a shock wave so forceful it left him paralyzed. Before he had time to compose himself, another shock wave rolled right through his body. A song of sorts, a voice so deep, it was as if the estuary bottom was cracking open beneath him, sucking him into an abysmal void. A voice so commanding it seemed to blow his river cosmos wide open, warping his perception, expanding the space around him, rolling through him, rattling his bones, mangling his flesh, turning him up and down and inside out. *Blue whale!* The voice boomed on and on in the endless waters all around, and there was no hiding, no refuge, no returning to the safe confines of the river.

The sky darkened. A shadow crept across the ocean. The temperature in the air above dropped. A wind was now rippling the surface. Other smolts arrived. Among them was one rather ugly smolt named Laxboy. He looked distastefully naked, outright ungainly. He didn't have fins but legs and arms. He didn't have scales but an unpleasantly swollen skin. "Not from around here," Desolax thought to himself. "Must be a distant cousin." He could see that Laxboy swam a little off to the side from everyone else. The wretched creature looked dazed, stupefied. Desolax swam up to him: "What's the matter with you?" he asked. Laxboy looked right past him, out toward the blue endlessness beyond the estuary. When he spoke, his voice nearly choked: "The eternal noise of these infinite spaces frightens me." Desolax followed his gaze. He thought: "What does it all mean, being salmon here on the perilous brink of this frightening ocean cosmos, in plain view of the infinite spaces beyond?" He couldn't deny it: Terror had crept into his cold blood, too. He felt as appalled by it all as that wretched cousin looked. But he also felt joyful, ecstatic even!

"Listen, Laxboy. I'll be straight with you. Right now I feel rather uncertain about just about everything. Except perhaps for the fact that my bones are still rattling from that whale cry. In a word, I'm terrified. There's no denying it. This great river seems too big and boundless and bottomless for my taste. But I also feel strangely alive this morning! My senses are stretching out as if after a good sleep; I feel as if my entire being is aching to radiate outward. I can't help sniffing this way and that; my eyes want to penetrate the deep blue; all along my lateral line I'm brushed by the strangest sounds, waves that break against distant cliffs, kelp forests that sway in a surge, air that bubbles from the deep, seals who chase each other. The point is . . . You seem a little lonely. Why don't you come along and we'll figure this out together? Keep close." So it went.

As the crescent moon was sinking into the ocean, they left the estuary. Two tiny smolts swam out into the vast unknown.

"To be thinking like the ocean. To be the ocean thinking itself within," Desolax thought one night as they were circling a large ocean eddy. "Perhaps that's what it means to reach maturity." Months had passed since they left the estuary. They had seen herds of whales, seabirds by the tens of thousands, and squid that seemed larger than life. Once, they had barely escaped an underwater volcanic eruption. Often, at dusk, they had watched in awe as unbelievable masses of zooplankton migrated up from the deep to their surface pastures—the largest animal mass migration on the planet, a huge diurnal sigh, unmatched even by his own kinfolks' mass migration. Somewhere along the line he had understood and accepted that his sentient body bore within itself the promise for metamorphosis, a creative adaptability within a world that never rests. He had grown and grown, and he had sensed inside of himself distant echoes of his birth river: its velocity, its seasonal temperaments, the power of its autumn swells, the complex topography of its arteries. Each quality of that distant watershed seemed to add its subtle claim to the growing body that was his, refining his sentient flesh. Born into a shapeshifting world, that's what he was: a shapeshifter. The Gulf Stream, the clouds that rose from algae blooms, the winter storms they had endured: Each resonated within his flesh, each demanded of him to seek a poetic attunement, a response.

There were moments still. Every now and then the old terror swept through him again, as dreadful as on that first morning outside their river cosmos. But he had begun to trust his body's poetic attunement with the ocean. There seemed to be wisdom in his body that was so deep, his mind could not fathom it. Desolax knew that staying inside the old river's wombish cosmos had never been a choice. The only choice was to accept and to get on with it.

Now his mind was in him and it was in the ocean, so much so that the ocean did, indeed, think itself within him. Each encounter with blackfish, tuna, or beluga bellowed and bowled through the water and jolted through his body, inciting him. Each spark set some region of his awareness ablaze. At times the sparks came flashing in such rapid torrents, from every angle, from near and far, that every fiber within him felt utterly irradiated. His sentience attuned itself instantly, effortlessly. He was awake, and he was alert. Navigating this formerly vast unknown had become a creative dance between perception, the memory of his breathing flesh, and the larger bodies that composed him. It had become a continuous improvisation with the upsurge of the present moment, an increasingly subtle and capable alertness to the real.

He also knew that his expanding experience of the real was not the actual tragedy, dreadful though it first had seemed. The actual tragedy was swimming beside him. Laxboy looked as miserable now as he had on that first morning. With the help of Desolax he'd made it this far, alright. He hadn't been eaten, unlike poor Galilax who'd been torn apart by a feverish shark one morning not long after the winter solstice. But wherever they swam, it somehow seemed to Desolax as if Laxboy just wasn't all there. In his dream, he cursed the great expanses. More often than not, he looked lost, disoriented. Once he had swum right up to a killer whale mother who was nurturing her calf. "I dare you," he had bellowed. "I dare you try and catch me. You won't! Unlike you, I think, therefore I am going to outsmart you." Just then, Desolax had rammed him in the side so hard that it left a bruise. Better that, Desolax had thought, than for the wretched cousin to end up as orca meat. "Come on, we need to get away from here, *now!*"

Desolax couldn't help but think that Laxboy had gone a little mad.

———

Accepting the responsibility of return is never a trivial matter. But every thought and every sentiment inside of him was craving closure. Every fiber of his embodied mind was bending itself back toward origins. As he had grown from a smolt into a mature fish, his sense for the larger ocean body had also grown keener. Each local variation in the blue expanse—its own field quality. Each region in the ocean—its particular magnetic tension. He knew, long before they even reached the continental shelf again: He knew that this vast, round river into which they had swum, back when they were just adolescent smolts, was also part of their wombish cosmos. He knew that they had never left the womb that was the great, round river. Now as ever, the ocean stretched out beneath the sky. Now as ever, the sky hung under the moon.

"I'm not sure whether I can make it all the way back to the river," Laxboy said to him one evening as they were riding a fast, cold undercurrent. Desolax kept on swimming straight ahead. He said: "We've both come far. We'd be fools to stop now." Laxboy caught up with him. He panted: "I'm just not sure whether I can." Desolax turned abruptly. He looked Laxboy straight in the eye: "Do this with me. We must complete the cycle. We must return to where we left back when. We must do it for one another, and we must do it for our own sake." And so it went.

A pulse came rolling toward them from the land, like the rhythmic calling of drums. It was the pulse of the dying season. Autumn rains made the estuary swell. Once they got close to the river, they instantly recognized that this was the one. It tasted like no other. They both felt an urge now. And so they swam, and they leapt, and they rested, and they swam some more, and every gurgling and splashing and gushing and sloshing here was familiar to their sentient bodies. The river song hadn't changed at all. And yet, nothing was as it had been. Back then, their sense of being had been as narrow as the river's horizon. Now they were both large, they contained multitudes. They were insides within insides within insides.

Journeys within journeys within journeys: From as deep as any two smolts had as yet ventured into the black Arctic Ocean—from as most definitely *outside* as any young salmon had ever been from their birth river—they had done as their ancestors always had struggled to do: They had endured the uncertainty, they had lived, and they had let themselves be drawn back again. They had accepted that returning home had never been their choice to make. It had been their responsibility.

Desolax was dying. His muscles were peeling off his bones. He was lying flat-sided on the riverbank; he could barely breathe. His flesh reeked of death. A fly crawled into a festering belly cavity. "Will you do something for me?" he asked Laxboy, who was kneeling by his side. "If you ever get to spawn, as I have, promise me that you'll do as all of us have always done. Come back right here and entrust your sons and daughters to the ancient waters."

"I promise, grandfather," Laxboy said. Then they were quiet together for a long time.

It was over. Laxboy stood up, stretched his limbs, and took one last look at the carcass by his feet. Two more flies were now buzzing round and round that putrid belly cavity. Laxboy nodded. Then he turned and scrambled up the riverbank. It was time to return to his village. At last he disappeared amid a thicket of black alders.

ACKNOWLEDGMENTS

*T*he poet Stephanie Mills once hosted a week-long writing retreat on the theme of "Writing from the Core of the Earth" at Schumacher College in Devon, England. As our small assembly of eager writers sat gathered in the opening circle on the first afternoon, Stephanie addressed us with words that were astonishingly insightful and elegant: "In writing there are no separate selves." Planning, researching, and writing this work has taken four years—five, if I count in the year I spent on paternity leave—and throughout this time I have had the good fortune of encountering others who gifted me with their kind attention, their shared enthusiasm, their respectful disagreement, and their friendship, and without whom this book would not exist. I wish to give my sincere thanks to them.

The book began as a PhD dissertation. Only gradually did I understand that I had made one of the most important choices in the course of writing it before I had penned even one word: in choosing my *Doktorvater*, Arne Johan Vetlesen. With his inimitable mix of sincerity, decency, humor, and a quickness to call a spade a spade, Vetlesen carried my work through several critical bottlenecks. Your guidance emboldened me; your example challenged me. Thank you for offering your unconditional guidance, for your perseverance, and for having my back. The gift of your writings, Bergljot Børresen, was an early inspiration. Our conversations began as formal interviews but quickly outgrew the formal setting in depth and richness. I feel privileged to have been able to know you throughout these years. Thank you for your friendship, your companionship, and your guidance at all stages of this work, from the first tentative writing sketches to the times when you sat down to critically comment on the near-finished draft. Kathy Fox, Douglas McDonnell, and Leslie McDonnell shared their home and guided me deep into the storied heart of the Pacific Northwest. Meeting you was pivotal. It was also a great joy. I think often of our long talks, of the woods behind your house, of the books we shared, the lavender cookies, Puget Sound, Fox Island, of your ravaged lake, of Rosie. Dear, good people!

Of the Lower Elwha Klallam Tribe on the mouth of the Elwha River, a special thanks to LaTrisha Ollom-Suggs, Anthony Fernandez, Jamie Valadez, and to the late Adeline Smith. Thank you also to Timothy Montler of the University of Texas for giving me the chance to witness your work on recording and restoring the Klallam language. To Marielle and Bart Eykemans of Port Angeles, I am indebted for your extraordinary hospitality. I was practically a stranger when you spontaneously invited me for a warm meal to your beautiful home on the eastern bank of the Elwha, and when on that first evening you suggested that my companion and I should live with you for the remainder of our visit to the Olympic Peninsula. I hope to be able to visit you again one day and take a walk by the Elwha with you, now that the salmon are returning. I also hope all the maple trees around your home are thriving. You, too, opened your home in Seattle, Joshua, Celeste, and Gary Cranston, and blessed me with your kindness. Alan Drengson and Tory Stevens graciously invited me for a visit to Vancouver Island. I appreciated the hike to Mt. Douglas and the visit to the Royal BC Museum, and I am particularly grateful for the books. Chuck Larsen and his wife, Jan, generously let me glimpse some of the riches of the Coast Salish oral tradition. I understood then that a story shared is not only a token of trust, but also a responsibility bestowed: to treat the story with respect and care. The writers Waziyakawin and Derrick Jensen each sent words of encouragement and offered some direction when I was first surveying the landscape of Native American cultures along the Pacific Rim. Polly Higgins opened doors of which neither of us knew at the time how important they would turn out to be for me. Our conversations challenged, nurtured, and thrilled me. Tell me, Polly, what makes your heart sing? I pray you'll succeed in your work. Satish Kumar, Paul Wapner, Lawrence Buell, Thomas Dunklin, Lynda Mapes, Georgiana Keable, Malcom Green, Heike Vester, Inge Sandven, Kimberly Eriksen, Erling Christiansen, Pablo Pachilla, Christian and Eduardo Tiscornia, Dag Skarstein, Arne Løyning, Torgeir Vassvik, Andreas Daugstad Leonardsen, Aage Solbakk, Niilas Somby, Dan Lewis, Bonny Lynn Glamback, and Charles Eisenstein have each been conversation partners at one point and helped me deepen my understanding of particulars I was chewing on at the time. Giovanni Roversi kindly helped me out with my poor Latin.

Konrad Ott, my first teacher in ecophilosophy, was the earliest of a number of extraordinary mentors. You were a patient guide at a time when I struggled even to ask relevant questions. Arne Næss was of singular importance early on in my development. Your writings fully redirected my path. Your youthful spirit and the ease with which you let yourself be seized by wonder are gifts I seek to honor through my own work. James Lovelock, too, was an early teacher, and your work, too, has been transformative. David Abram, during his tenure as Arne Næss Professor of Social Justice and Ecology at the University of Oslo, graciously agreed to co-teach a graduate course with me. Our joint explorations both within the course and without have enriched this work in so many ways. Thank you for having been a generous and kind mentor to me. To Ingrid Stefanovic and Stephan Harding, my sincere thanks for your attentiveness and your encouragement in yet another critical moment. Thank you also for helping this book find its way to Chelsea Green. To Shaun Chamberlin I am grateful for championing the work. Margo Baldwin and Brianne Goodspeed have given such reassurance and graceful advice that I never doubted the book was in good hands. You two represent to me the good heart of the publishing world, and I admire the work you do. Rory Bradley: From your first review of the book draft to the final edits you suggested, I have felt such gratitude and joy in knowing you were my editor. It has been my distinct privilege to learn from you, to laugh with you, to witness your style and skill and elegance, and to see how gradually, with the guidance of your capable hands, the pages began to sing.

The Center for Development and the Environment (SUM) at the University of Oslo provided me with good structures to go about my daily work. Nina Witoszek was the first to endorse an early sketch of my work. Maren Aase, Ulrikke Wethal, Thorunn Gullaksen Endreson, Arve Hansen, and Desmond McNeill were attentive peers in our work-in-progress seminars. Your comments, even your resistance, were valuable. Kristoffer Ring was the one person I knew I would be able to fall back on when my computer failed me. But more than that, your cheerful presence graced the monotonous physical structure of the offices. The same is true for Anne-Line Sandåker, Hilde Holsten, Manhar Pat Harmansen, Terje Røysum, and Monica Guillén-Royo, as well as for Sumia, the kind cleaning lady at SUM. My graduate students of the spring course Philosophical Foundations of Ecomodernity at the University of Oslo, as well as my undergraduate students in Buenos Aires, have inspired me through your wittiness, your thirst to learn, your eagerness to encounter the more-than-human world in its ineffable richness, and your willingness to let yourselves be vulnerable and compassionate. In times as grim as these, working with you continues to imbue me with hope. Mickey Gjerris and Helena Røcklingsberg arranged and led me through a week of intense and fine learning in the remote village of Kabelvåg, on Lofoten Islands. Together with the fellow participants of the Contemplative Environmental Studies Retreat on Lama Mountain, New Mexico, I delved into a delightfully odd exploration of the fluid and participatory relationship between mind and body, intellect and art, despair and courage.

Anita Liebenau, Ernesto Adrian Lara, Jake McDougald, Annelie Ott, Siri Kalla, Lena Groß, and Kit-Fai Næss have been dear and steadfast friends. Each of you has sustained me and my work in ways that make you unique and beautiful. The same is true for Tiril Bryn, who kindly agreed to read the manuscript when I thought it near-finished. You showed me it wasn't and gently drew me into a spiraling series of shared explorations that opened up further aspects of my work, aspects I would not have discovered without the enthusiasm and creativity of such a gifted storyteller as yourself. I remember vividly the many walks and conversations we had, Johanna Becker. Many early ideas of this work took shape at the leisurely pace of our feet, by lakesides, in alpine mountains, on forested hilltops, in cities; thank you for your thoughtfulness and your care. Anders, Kari, Aurora, and Alma Baumberger reminded me time and again (gently but persistently) that wildness dwells just beyond our day-to-day attention. For the nights in the snow at twenty-five degrees below Celsius, in the mountain cabin in Rondane, in the old fire watchtower overlooking the Østmarka forest, in the marshes awaiting the elusive black grouse's mating dance, skiing off-trail in search of wolf tracks: for all those nights, thank you. I am sorry for the many other nights I have missed out on. Øyvind and Molly Eliason shared drinking water at the coldest stretch of winter, a bike pump in spring, herbs for tea and cooking, neighborly companionship, and stories. My family nourished and supported me despite the physical distance between us. Thank you for being a part of me, Ian Micah, Malina Jewel, Christopher Lee and Romy McNelly, Anna-Teresa Rudick, Siegmar and Manuela Müller, Frank Habelt, Isolde and Lothar Köhn. To my dear girls, Katrin and Kaia Isolde Vels, my sincere thanks for bearing so generously with me when I struggled to help the manuscript fledge. You two are my home and my harbor, and I am thankful we're in this together. Kati, you are my partner and accomplice, first critic and first supporter, fellow animal-lover and fellow voyager through this weird, bumpy, imperfect, precious thing which is life.

To all, I hope that these pages do justice to the kindness you have given so freely and abundantly.

But my gratitude does not end here. My intimate awareness of being a relational self extends beyond these dear human companions. For there are others who have also enriched me on this journey, in ways subtle and not always easy to bring to the level of conscious thought. There are the salmon of the many streams I visited along the way: among them the Akersevla, Ljanselva, Daleelva, Vossoelva, Altaelva, Tanaelva, the Elwha, Klamath, Columbia, Fraser, the Rhine, and, not least, the watershed of my birth and upbringing, the Oder on the border between Poland and Germany. I was well into my thirties when I learned, with some amazement, that I, too, had in fact grown up alongside a once-thriving salmon river! Beside the salmon and the rivers, I recognize the subtle and steady gift-giving of others, too: old spruce outside the rundown cabin where I lived for some time, mother of twenty-some saplings, all connected to her maternal stem through delicate and implausibly long, wooden umbilical branches; the forest itself, gifting me with breath, snow, flower bloom, drinking water, cloudberries, blueberries, lingonberries, forest strawberries, raspberries, with dreams, sunshine, shadow, with chanterelle, hawk's wing, birch bolete, porcini, with the skills to fell a tree and the gratitude for the life it has sacrificed so as to keep me warm during Norway's long winters. My thanks also to my animal neighbors who gifted me with their presence: red fox, squirrel, beaver, moose, roe deer, mouse weasel, mouse (ever mischievous), invisible wolf (leaving but tracks on the frozen lake outside the cabin), secretive black grouse, wood grouse, watchful jay (leaving no doubt as to who was really in charge on the forest clearing where I lived), magpie, crow, Canada goose, nuthatch, grey goose, mallard, bullfinch, coal tit, blue tit, pileated woodpecker, green woodpecker, black woodpecker, grey heron, common crane, blindworm, grass snake, adder, and, not least, Tuti, dear companion.

I extend a final token of gratitude to the green ones who have given their flesh to become the paper on which these pages are printed.

BIBLIOGRAPHY

Aas, Lars Reinholt. "Våre Fjerne Slektniger i Himalaya." *Aftenposten*, August 20, 2014. Accessed August 20, 2014, http://www.aftenposten.no/viten/Vare-fjerne -slektninger-i-Himalaya-7614522.html.

Abram, David. *Becoming Animal: An Earthly Cosmology.* New York: Vintage, 2010.

———. "In the Depths of a Breathing Planet: Gaia and the Transformation of Experience." In *Gaia in Turmoil: Climate Change, Biodepletion, and Earth Ethics in an Age of Crisis*, edited by Eileen Christ and H. Bruce Rinker, 221–242. Cambridge, MA: MIT Press, 2010.

———. "The Mechanical and the Organic: On the Impact of Metaphor in Science." In *Scientists on Gaia*, edited by Stephen Schneider and Penelope Boston. Boston, MA: MIT Press. Retrieved from Alliance for Wild Ethics, May 9, 2017, http://wildethics.org/essay/the-mechanical-and-the-organic/.

———. "Reciprocity." In *Rethinking Nature: Essays in Environmental Philosophy*, edited by Bruce V. Foltz and Robert Frodeman, 77–92. Bloomington: Indiana University Press, 2004.

———. *The Spell of the Sensuous.* New York: Random House, 1996.

———. "Storytelling and Wonder: On the Rejuvenation of Oral Culture." Retrieved from Alliance for Wild Ethics, May 9, 2017, http://wildethics.org /essay/storytelling-and-wonder/.

Abram, David, and Derrick Jensen. "David Abram Interviewed by Derrick Jensen." In *How Shall I Live My Life: On Liberating the Earth from Civilization*, edited by Derrick Jensen. Oakland, CA: PM Press, 2008. Retrieved from Alliance for Wild Ethics, May 9, 2017, http://wildethics.org/essay/david-abram -interviewed-by-derrick-jensen/.

Adamson, Thelma, ed. *Folk Tales of the Coast Salish.* Lincoln: University of Nebraska, 1934/2009.

Aldwell, Thomas. *Conquering the Last Frontier.* Seattle: Artcraft Engraving and Electrolyte Company, 1950.

Allenby, Braden R., and Daniel Sarewitz. *The Techno-Human Condition.* Cambridge, MA: MIT Press, 2011.

Alliance for Natural Health. "Action Alert: Genetically Modified Frankenfish!" August 31, 2010. Accessed April 19, 2012, http://www.anh-usa.org/action -alert-genetically-modified-frankenfish/.

Amos, John "Bottom Trawling: Sediment Plumes Visible From Space." February 15, 2008. Accessed May 9, 2017, http://blog.skytruth.org/2008/02/bottom-trawling-sediment-plumes-visible.html.

AquaBounty Technologies, Inc. "AquAdvantage Fish." Accessed March 1, 2014, http://www.aquabounty.com/products/products-295.aspx.

———. "AquaBounty Fact Sheet." Accessed March 1, 2014, http://www.aquabounty.com/documents/press/2010/AquaBounty%20Fact%20Sheet.pdf.

———. "Environmental Assessment for AquAdvantage® Salmon." August 25, 2010. Accessed March 1, 2014, https://cban.ca/wp-content/uploads/AAS_EA-redacted.pdf.

Aquinas, Thomas. "Quaestiones Disputatae de Veritate." Corpus Thomisticum, 1256–1259. http://www.corpusthomisticum.org/.

Aristotle. *Metaphysics*. Edited by Gregory R. Crane. Perseus Digital Library. Accessed March 25, 2014, http://www.perseus.tufts.edu/.

Bacon, Francis. "Aphorisms Concerning the Interpretation of Nature and the Kingdom of Man." In *The Philosophical Works of Francis Bacon*, edited by John M. Robertson. London: Routledge, 2012.

———. *The New Atlantis*. 1623/27. Accessed March 20, 2014, http://www.fcsh.unl.pt/docentes/rmonteiro/pdf/The_New_Atlantis.pdf.

Barbarić, Damir. "Die Grenze zum Unsagbaren. Sprache als Horizont Einer Hermeneutischen Ontologie (GW 1, 442-478)." In *Klassiker Auslegen. Hans-Georg Gadamer: Wahrheit und Methode*, edited by Günter Figal, 199–218. Berlin: Akademie Verlag, 2007.

Barnosky, Anthony D., Nicholas Matzke, Susumu Tomiya, Guinevere O. U. Wogan, Brian Swartz, Tiago B. Quental, Charles Marshal, Jenny L. McGuire, Emily L. Lindsey, Kaitlin C. Maguire, Ben Mersey, and Elizabeth A. Ferrer. "Has the Earth's Sixth Mass Extinction Already Arrived?" *Nature*, vol. 471, no. 7336 (2011): 51–57. doi:10.1038/nature09678.

Barnosky, Anthony D., Paul L. Koch, Robert S. Feranec, Scott L. Wing, and Alan B. Shabel. "Assessing the Causes of Late Pleistocene Extinctions on the Continents." *Science*, vol. 306, no. 5693 (2004): 70–75. doi:10.1126/science.1101476.

Beardmore, John A., and Joanne S. Porter. "Genetically Modified Organisms and Aquaculture." *FAO Fisheries Circular*, no. 989 (2003). Accessed February 15, 2014, ftp://ftp.fao.org/docrep/fao/006/y4955e/Y4955E00.pdf.

Berge, Aslak. "Disse Artene Leder den Blå Revolusjonen." June 17, 2014. http://www.ilaks.no/disse-artene-leder-den-bla-revolusjonen/.

Berg-Hansen, Lisbeth. "Laks på Land. Å Flytte Norsk Lakseoppdrett på Land er Urealistisk, det er Heller Ikke Oproblematisk." *Dagbladet*, October 11, 2012. http://www.agbladet.no/2012/10/11/kultur/debatt/debattinnlegg/fiskeri/laks/23825741/.

Berry, Thomas. *The Dream of the Earth.* San Francisco: Sierra Club Books, 1998.
———. *The Story of the Universe.* San Francisco: Harper, 1992.
Berry, Wendell. *Life Is a Miracle: An Essay Against Modern Superstition.* Berkeley, CA: Counterpoint, 2001.
Betts, Richard A. "Offset of the Potential Carbon Sink from Boreal Forestation by Decreases in Surface Albedo." *Nature,* vol. 408, no. 6809 (2000): 187–190.
Børresen, Bergljot. *Fiskenes Ukjente Liv.* Oslo: Transit, 2007.
Boyd, C. E. "Change Is Coming: History, Identity and the Land among the Lower Elwha Klallam Tribe of the North Olympic Peninsula." PhD diss., University of Washington, 2001.
Brezsny, Rob. *Pronoia Is the Antidote for Paranoia.* Berkeley, CA: North Atlantic Books, 2009.
Brown, Alice. "Descartes' Dream." *Journal of the Warburg and Courtauld Institutes,* vol. 40 (1977): 256–273. Accessed November, 16, 2011, http://www.jstor.org/stable/750999.
Brown, Lester. "Plan B Updates: Paving the Planet: Cars and Crops Compete for Land." Earth Policy Institute, February 14, 2001. Accessed May 14, 2013, http://www.earth-policy.org/plan_b_updates/2000/alert12.
Brunner, Eric, Erik Millstone, and Sue Mayer. "Beyond 'Substantial equivalence'." *Nature,* vol. 401, no. 6753 (1999): 525–526.
Buhner, Stephen Harrod. *The Lost Language of Plants: The Ecological Importance of Plant Medicines of Life on Earth.* White River Junction, VT: Chelsea Green, 2002.
Bump, Joseph K., Rolf O. Peterson, and John A. Vucetich. "Wolves Modulate Soil Nutrient Heterogeneity and Foliar Nitrogen by Configuring the Distribution of Ungulate Carcasses." *Ecology,* vol. 90, no. 11 (2009): 3159–3167. http://dx.doi.org/10.1890/09-0292.1.
Burgess, Seth D., Samuel Bowring, and Shu-zhong Shen. "High-Precision Timeline for Earth's Most Severe Extinction." *Proceedings of the National Academy of Sciences of the United States of America.* January 2, 2014. Accessed March 4, 2015, http://www.pnas.org/content/111/9/3316.abstract.
Callicott, J. Baird. *Earth's Insights: A Multicultural Survey of Ecological Ethics from the Mediterranean Basin to the Australian Outback.* Berkeley: University of California Press, 1994.
———. *In Defense of the Land Ethic.* Albany, NY: State University of New York Press, 1989.
Campbell, Sarah K., and Virginia L. Butler. "Archaeological Evidence for Resilience of Pacific Northwest Salmon Populations and the Socioecological System over the Last ~7,500 Years." *Ecology and Society,* vol. 15, no. 1 (2010): 17. Accessed December 3, 2014, http://www.ecologyandsociety.org/vol15/iss1/art17/.
Capra, Fritjof. *The Tao of Physics.* Toronto: Bantam Books, 1975.

Capra, Fritjof, and Pier Luigi Luisi. *The Systems View of Life*. Cambridge, UK: Cambridge University Press, 2015.

Carson, Rachel L. *The Sea Around Us*. Oxford: Oxford University Press, 1989.

———. *Silent Spring*. Boston: Houghton Mifflin Company, 1962/1987.

Catlin, George. "Letters and Notes." 1841. Accessed May 6, 2014, http://americanart .si.edu/exhibitions/online/catlinclassroom/catlin_browsec.cfm?id=413.

Cederholm, C. J., D. B. Houston, D. L. Cole, and W. J. Scarlett. "Fate of Coho Salmon *(Oncorhynchus kisutch)* Carcasses in Spawning Streams." *Canadian Journal of Fisheries and Aquatic Sciences*, vol. 46, no. 8 (1989): 1347–1355. doi:10.1139/f89-173.

Charles, Benjamin. *Klallam Stories*. Edited by Jamie R. Valadez. Lower Elwha Klallam Tribe, 2012.

Charlson, Robert J., James Lovelock, Meinrat O. Andreae, and Stephen G. Warren. "Oceanic Phytoplankton, Atmospheric Sulphur, Cloud Albedo and Climate." *Nature*, vol. 326, no. 6114 (1987): 655–661.

Chopra, Aditya, and Charles H. Lineweaver. "The Case for a Gaian Bottleneck: The Biology of Habitability." *Astrobiology*, vol. 16, no. 1 (2016). doi:10.1089 /ast.2015.1387.

Christ, Eileen. "Intimations of Gaia." In *Gaia in Turmoil: Climate Change, Biodepletion, and Earth Ethics in an Age of Crisis*, edited by Eileen Christ and H. Bruce Rinker, 315–333. Cambridge, MA: MIT Press, 2010.

Christie, Douglas E. *The Blue Sapphire of the Mind: Notes for a Contemplative Ecology*. Oxford: Oxford University Press, 2013.

Clark, Ella E. *Indian Legends of the Pacific Northwest*. Berkeley, CA: University of California Press, 1984.

Clayoquot Action. "Clayoquot Action." Accessed January 1, 2016, http:// clayoquotaction.org/.

Clutton-Brock, Juliet. *A Natural History of Domesticated Mammals*. Cambridge, UK: Cambridge University Press, 1987/1989.

———, ed. *The Walking Larder: Patterns of Domestication, Pastoralism, and Predation*. London: Unwin Hyman, 1990.

Coates, Peter. *Salmon*. Chicago: University of Chicago Press, 2006.

Cohen, H. Floris. *The Scientific Revolution: A Historiographical Inquiry*. Chicago: University of Chicago Press, 1994.

Colombi, Benedict J., and James F. Brooks, eds. *Keystone Nations: Indigenous Peoples and Salmon across the North Pacific*. Santa Fe, NM: School for Advanced Research Press, 2012.

Crane, Jeff. *Finding the River: An Environmental History of the Elwha*. Portland: Oregon State University Press, 2011.

Curry, Patrick. *Ecological Ethics: An Introduction*. Cambridge, UK: Polity Press, 2011.

Delfinschutz. "Delfinschutz." Accessed October 26, 2012, http://www
.delphinschutz.org/wissen/fischerei/fischkonsum_deutschland.html.

Deloria, Vine. *God Is Red: A Native View of Religion*. Golden, CO: Fulcrum
Publishing, 2003.

———. *Indians of the Pacific Northwest: From the Coming of the White Man to the
Present Day*. Golden, CO: Fulcrum Publishing, 2012.

Descartes, René. *Discourse on Method for Reasoning Well and for Seeking Truth in
the Sciences*. Translated by Ian Johnston. 1637. Accessed February 17, 2014,
https://www.oneeyedman.net/school-archive/classes/fulltext/www.mala.bc.ca
/~johnstoi/descartes/descartes1.htm.

———. "Meditations." In *The Philosophical Works of Descartes*, vol. 1, translated
by Elizabeth Sanderson Haldane. Cambridge, UK: Cambridge University
Press, 1968.

Dickinson, Emily. *The Complete Poems of Emily Dickinson*. Edited by Thomas H.
Johnson. Boston: Little, Brown, and Company, 1961.

Dolce, Chris. "Pacific Northwest Snowpack Near Lowest Levels on Record." The
Weather Channel, February 11, 2015. Accessed December 17, 2015, http://
www.weather.com/news/news/snowpack-cascades-west-northwest.

Doney, Scott C., Victoria J. Fabry, Richard A. Feely, and Joan A. Kleypas. "Ocean
Acidification: The Other CO_2 Problem." *Annual Review of Marine Science*,
vol. 1 (2009): 169–192. doi:10.1146/annurev.marine.010908.163834.

Donne, John. "LXXX Sermons Preached by That Learned and Reverend Divine
Iohn Donne, Dr. in Divinity, Late Deane of the Cathedrall Church of S. Pauls
London." 1630/1997. Accessed October 26, 2013, http://www.ccel.org/d
/donne/sermons/easter.html.

Doughty, C. E., J. Roman, S. Faurby, A. Wolf, A. Haque, E. S. Bakker, Y. Malhi,
J. Dunning, and J. C. Svenning. "Global Nutrient Transport in a World of
Giants," *Proceedings of the National Academy of Sciences of the United States of
America*, vol. 113, no. 4 (October 2015): 868–873.

Drake, Stillman. *Galileo at Work: His Scientific Biography*. Chicago: University of
Chicago Press, 1978.

Drengson, Alan R. *Wild Way Home*. Victoria, BC: Light Star Press, 2010.

Duquette, David A., ed. *Hegel's History of Philosophy: New Interpretations*. Albany,
NY: State University of New York Press, 2003.

Egan, Virginia G. "Restoring the Elwha: Salmon, Dams and People on the
Olympic Peninsula; A Case Study of Environmental Decision-Making." PhD
diss., Antioch University of New England, 2007.

Egeland, John Olav. "Nå Dreper de Fjordene." *Dagbladet*, June 8, 2015. http://
www.dagbladet.no/2015/06/08/kultur/meninger/hovedkommentar
/kommentar/fiskeoppdrett/39555406/.

Eisenstein, Charles. *The More Beautiful World Our Hearts Know Is Possible*. Berkeley, CA: North Atlantic Books, 2013.

Emerson, Ralph Emerson. "The Poet." 1844. Retrieved from Poetry Foundation, January 19, 2014, http://www.poetryfoundation.org/learning/essay/237846 ?page=7.

Ettinger, Barbara. *A Sea Change*. Directed by Barbara Ettinger. Niijii Films, 2009. Accessed October 1, 2014, http://www.aseachange.net/.

Evernden, Neil. *The Natural Alien: Humankind and Environment*. Toronto: University of Toronto Press, 1985.

Fabry, Victoria J., Brad A. Seibel, Richard A. Feely, and James Orr. "Impacts of Ocean Acidification on Marine Fauna and Ecosystem Processes." *ICES Journal of Marine Science*, vol. 65 (2008): 414–432. doi:10.1093/icesjms/fsn048.

Feely, Richard A., Christopher L. Sabine, and Victoria J. Fabry. "Carbon Dioxide and Our Ocean Legacy." April 2006. Accessed October 26, 2011, http://www.pmel.noaa.gov/pubs/PDF/feel2899/feel2899.pdf.

Fisher, Andy. *Radical Ecopsychology: Psychology in the Service of Life*. Albany: State University of New York Press, 2002.

Foltz, Bruce. *Inhabiting Earth: Heidegger, Environmental Ethics, and the Metaphysics of Nature*. New Jersey: Humanities Press, 1995.

Foster, James. "Salmon Successfully Farmed Inland." February 24, 2012. Accessed April 16, 2012, http://responsibleaquaculture.wordpress.com/2012/02/24 /salmon-successfully-farmed-inland/.

Fromm, Erich. "Marx's Concept of Socialism." 1961. Accessed October 30, 2014, https://www.marxists.org/archive/fromm/works/1961/man/ch06.htm.

Gadamer, Hans-Georg. *Hermeneutik I. Wahrheit und Methode. Grundzüge einer Hermeneutik. GW 1*. Mohr Siebeck: Tübingen, 1986.

———. *Truth and Method*. New York: Continuum, 1997.

Gagliano, Monica, Stefano Mancuso, and Daniel Robert. "Towards Understanding Plant Bioacoustics." *Trends in Plant Science*, vol. 17 no. 6 (2012): 323–325. doi:10.1016/j.tplants.2012.03.002.

Galilei, Galileo. *The Assayer*. Translated by Stillman Drake. In *Discoveries and Opinions of Galileo* (New York: Doubleday & Co., 1957). Retrieved from Princeton University, July 1, 2015,http://www.princeton.edu/~hos/h291/assayer.htm.

Gander, Hans-Helmuth. "Erhebung der Geschichtlichkeit des Verstehens zum Hermeneutischen Prinzip. (GW 1, 270–311)." In *Klassiker Auslegen. Hans-Georg Gadamer: Wahrheit und Methode*, edited by Günter Figal, 105–125. Berlin: Akademie Verlag, 2007.

Gåsvatn, Kari. "Sykehuset med Fabrikklogikk." *Nationen*. July 19, 2013. http://www.nationen.no/tunmedia/sykehus-med-fabrikklogikk/.

Geelmuyden, Niels Christian. *Sannheten på Bordet*. Oslo: Capellen Damm, 2013.

Geiling, Natasha. "Why Are Hundreds of Thousands of Salmon Dying in the Northwest?" *ClimateProgress*, vol. 29 (July 2015). Accessed July 29, 2015, http://thinkprogress.org/climate/2015/07/29/3685149/dead-salmon-pacific-northwest/.

Gende, Scott M., Richard T. Edwards, Mary F. Willson, and Mark S. Wipfli. "Pacific Salmon in Aquatic and Terrestrial Ecosystems: Pacific Salmon Subsidize Freshwater and Terrestrial Ecosystems through Several Pathways, Which Generates Unique Management and Conservation Issues but Also Provides Valuable Research Opportunities." *BioScience*, vol. 52, no. 10 (2002): 917–928. https://doi.org/10.1641/0006-3568(2002)052[0917:PSIAAT]2.0.CO;2.

Gill, Jacquelyn L., John W. Williams, Stephen T. Jackson, Katherine B. Lininger, and Guy S. Robinson. "Pleistocene Megafaunal Collapse, Novel Plant Communities, and Enhanced Fire Regimes in North America." *Science*, vol. 326, no. 5956 (2002): 1100–1103. doi:10.1126/science.1179504.

Glendinning, Chellis. *My Name Is Chellis, and I'm in Recovery from Industrial Civilization*. Gabriola Island, BC: Canada, 1994.

———. *Off the Map: An Expedition Deep into Empire and the Global Economy*. Gabriola Island, BC: New Society Publishers, 2002.

Goodman, T. "Whale Poop Is Vital for Ocean Ecology." Halcyon Solutions, Inc., 2010. Accessed March 25, 2012, http://inventorspot.com/articles/whale_poop_vital_ocean_ecology.

Gray, John. *Heresies: Against Progress and Other Illusions*. London: Granta Books, 99.

Greenberg, Paul. *Four Fish: The Future of the Last Wild Food*. New York: Penguin Press, 2010.

Gresh, Ted, Jim Lichatowich, and Peter Schoonmaker. *Salmon Decline Creates Nutrient Deficit in Northwest Streams*. Inforain.org, January 2000. Accessed March 24, 2012,http://www.inforain.org/reports/salmon_decline.html.

Grinnell, Elaine. *Grandma's Stories: Ancestral Stories of the Klallam People*. DVD. Jamestown S'Klallam Tribe, 2007.

Grober, Ulrich. *Die Entdeckung der Nachhaltigkeit. Kulturgeschichte eines Begriffs*. München: Verlag Antje Kunstmann, 2010.

Guignon, Charles B. *Heidegger and the Problem of Knowledge*. Indianapolis: Hackett Publishing Company, 1983.

Gunther, Erna. "An Analysis of the First Salmon Ceremony." *American Anthropologist*, vol. 28, no. 4 (1926): 605–617.

———. *Klallam Ethnography*. Seattle: University of Washington Press, 1927.

Haeberlin, Hermann, and Erna Gunther. *The Indians of Puget Sound*. Seattle: University of Washington Press, 1930/1973.

Hanh, Thich Nhat. *Love Letter to the Earth*. Berkeley, CA: Parallax Press, 2013.

Hannesson, Rögnvaldur. "Hva Skal Vi med Villaksen?" *Dagens Næringsliv*, August 26, 2010.

Harari, Yuval Noah. "Industrial Farming Is One of the Worst Crimes in History." *Guardian*, September 25, 2015. http://www.theguardian.com/books/2015/sep/25/industrial-farming-one-worst-crimes-history-ethical-question.

Harding, Stephan. *Animate Earth: Science, Intuition, and Gaia*. Totnes, Devon: Green Books Ltd., 2006.

Harding, Stephan, and Lynn Margulis. "Water Gaia: Three and a Half Thousand Million Years of Wetness on Planet Earth." In *Gaia in Turmoil*, edited by Eileen Crist and Bruce Rinker. Cambridge, MA: MIT Press, 2009. Retrieved from Alliance for Wild Ethics, July 1, 2014, http://wildethics.org/essay/water-gaia.

Heidegger, Martin. *Sein und Zeit*. 12th edition. Tuebingen: Max Neimeyer, 1927/1972.

———. *Was Ist Das—Die Philosophie?* Günther Neske: Pfullingen, 1956.

Heinz Awards. "Richard A. Feely." 2010. Accessed November 25, 2013, http://www.heinzawards.net/recipients/richard-feely.

Helfield, James, and Robert Naiman. "Effects of Salmon-Derived Nitrogen on Riparian Forest Growth and Implications for Stream Productivity." *Ecology*, vol. 82, no. 9 (2001): 2403–2409.

Hofstader, Dan. *The Earth Moves: Galileo and the Roman Inquisition*. New York: W.W. Norton and Company, 2009.

House, Freeman. *Totem Salmon*. Boston: Beacon Press, 1999.

Hunt, Ed. "137 Species Rely on Pacific Salmon." Salmon Nation. Accessed March 24, 2013, http://www.salmonnation.com/fish/137species.html.

Hyde, Lewis. *The Gift: How the Creative Spirit Transforms the World*. Edinburgh: Canongate Books Ltd., 1979/2006.

Jensen, Derrick. *Dreams*. New York: Seven Stories Press, 2011.

———. *Endgame*. New York: Seven Stories Press, 2004.

———. *How Shall I Live My Life? On Liberating the Earth from Civilization*. Oakland, CA: PM Press, 2008.

———. *A Language Older Than Words*. New York: Context Books, 2000.

———. *Listening to the Land: Conversations about Nature, Culture, and Eros*. White River Junction, VT: Chelsea Green Publishing, 2004.

———. *Thought to Exist in the Wild. Awakening from the Nightmare of Zoos*. Santa Cruz, CA: No Voice Unheard, 2007

Jensen, Derrick, Aric McBay, and Lierre Keith. *Deep Green Resistance*. New York: Seven Stories Press, 2011.

Johansson, Jan. *Laks*. Oslo: Landbruksforlaget, 1997.

Johnsen, D. Bruce. "Salmon, Science, and Reciprocity on the Northwest Coast." *Ecology and Society*, vol. 14, no. 2 (2009): 43. http://www.ecologyandsociety.org/vol14/iss2/art43/.

Kahane, Guy. "Our Cosmic Insignificance." *Noûs*, vol. 48, no. 8 (2014): 745–772. doi:10.1111/nous.12030.

Kearns, Peter, and Paul Mayers. "Substantial Equivalence Is a Useful Tool." *Nature*, vol. 401, no. 6754 (October 14, 1999): 640.

Keith, Lierre. *The Vegetarian Myth*. Crescent City: Flashpoint Press, 2009.

Kiplinger, Allysyn. "Groping Our Way toward a New Geologic Era: A Study of the Word Ecozoic." *Ecozoic Times* (Spring 2010). Accessed October 26, 2014, https://ecozoictimes.com/articles-2/groping-our-way-toward-a-new-geologic-era/.

Kristoffersen, Svein. "Vil Satse Bærekraftig." *Klassekampen*, February 18, 2011.

Krogshus, Mari. "Norsk Laks: Det Dødelige Pengeeventyret." Accessed February 11, 2015, http://www.improvinglives.no/norsk-laks-det-dodelige-pengeeventyret/.

Langberg, Øystein Kløvstad. "Kina Skal Produsere Laks i Ørkenen." *Aftenposten*, October 6, 2012. http://www.aftenposten.no/okonomi/Kina-skal-produsere -laks-i-orkenen-7010054.html.

Latour, Bruno. *Facing Gaia. Six Lectures on the Political Theology of Nature, Being the Gifford Lectures on Natural Religion*. February 18–23, 2013. Edinburgh: University of Edinburgh, 2013. Streams available on YouTube.

———. *We Have Never Been Modern*. Cambridge, MA: Harvard University Press, 1993.

Lavery, Trish J., Ben Roudnew, Peter Gill, Justin Seymour, Laurent Seuront, Genevieve Johnson, James G. Mitchell, and Victor Smetacek. "Iron Defecation by Sperm Whales Stimulates Carbon Export in the Southern Ocean." *Proceedings: Biological Sciences*, vol. 277, no. 1699 (2010): 3527–3531. doi:10.1098 /rspb.2010.0863.

Leopold, Aldo. *Round River: From the Journals of Aldo Leopold*. Oxford: Oxford University Press, 1993.

Lichatowich, Jim. *Salmon without Rivers: A History of the Pacific Salmon Crisis*. Washington, DC: Island Press, 1999.

Lien, Marianne E., and John Law. "Emergent Aliens: Performing Indigeneity and Other Ways of Doing Salmon in Norway." The Open University, March 10, 2010. Accessed March 20, 2012, http://www.sv.uio.no/sai/english/research /projects/newcomers/publications/working-papers-web/Emergent%20aliens %20Ethnos%20revised%20WP%20version.pdf.

Lorde, Audrey. "Poetry Is Not a Luxury." In *Sister Outsider: Essays and Speeches*. Berkeley, CA: Crossing Press, 1985. Retrieved from OnBeing, March 23, 2014, http://www.onbeing.org/program/words-shimmer/feature/poetry-not-luxury -audre-lorde/318.

Lovelock, James. *The Revenge of Gaia: Earth's Climate Crisis and the Fate of Humanity*. London: Penguin, 2006.

Magenau, Jörg. "Erwin Strittmatter—zum 100. Geburtstag." Süddeutsche Zeitung, August 15, 2012. http://www.getidan.de/gesellschaft/joerg_magenau/45862 /erwin-strittmatter-zum-100-geburtstag-14-08-1912.

Magnussøn, Anne. *Making Food: Enactment and Communication of Knowledge in Salmon Aquaculture*. Oslo: University of Oslo, 2011.

Mander, Jerry. *In the Absence of the Sacred*. San Francisco: Sierra Club Books, 1992.

Mapes, Lynda. "Back to Nature: Last Chunk of Elwha Dams Out in September." *Seattle Times*, August 16, 2014. http://www.seattletimes.com/seattle-news/back -to-nature-last-chunk-of-elwha-dams-out-in-september/.

———. *Breaking Ground: The Lower Elwha Klallam and the Unearthing of Tse-whit-zen Village*. Seattle: University of Washington Press, 2009.

———. *Elwha: A River Reborn*. Seattle: Seattle Press, 2013.

———. "Honoring First Salmon Caught below Elwha Dam—for the Last Time." *Seattle Times*, August 19, 2011. http://seattletimes.nwsource.com/html/fieldnotes /2015956009_honoring_first_salmon_caught_below_elwha_dam_-_for_the _last_time.html.

Maracle, Lee. "Where Love Winds Itself Around Desire." In *First Fish, First People: Salmon Tales of the North Pacific Rim*, edited by Judith Roche and Meg McHutchison, 160–179. Seattle: University of Washington Press, 1998.

Margulis, Lynn, and Dorion Sagan. *Microcosmos: Four Billion Years of Microbial Evolution*. Berkeley, CA: University of California Press, 1997.

———. *What Is Life?* New York: Simon and Schuster, 2000.

Martin, Paul Schultz. *Twilight of the Mammoths: Ice Age Extinctions and the Rewilding of America*. Berkeley, CA: University of California Press, 2005.

Martinsen, Siri. "Fisken, det Usynlige Dyret." *Aftenposten*, June 6, 2015. http:// www.aftenposten.no/meninger/kronikker/Veterinar-Siri-Martinsen-Fisken ---det-usynlige-dyret-8045843.html.

Masson, Jeffrey Moussaieff, and Susan McCarthy. *When Elephants Weep: The Emotional Lives of Animals*. New York: Delacorte Press, 1995.

Mathews, Freya. "Letting the World Grow Old: An Ethos of Countermodernity." In *Environmental Ethics: What Really Matters, What Really Works*, edited by David Schmidtz and Elizabeth Willott, 221–230. Oxford: Oxford University Press, 2002.

Mathismoen, Ole. "Om 25 År Kan Alt Være Borte." *Aftenposten*, September 13, 2014. http://www.aftenposten.no/nyheter/uriks/Om-25-ar-kan-alt-vare-borte -7704036.html.

McConnell, James. "Whittaker's Correlation of Physics and Philosophy." *Proceedings of the Edinburgh Mathematical Society*, vol. 11, no. 1 (1958): 57–68. Accessed October 31, 2014, http://www-history.mcs.st-andrews.ac.uk/Extras /Whittaker_physics.html.

Meadows, Robin. "A Good Fish for the Wine." *Conservation Magazine*, July 2008. Accessed March 30, 2012, http://www.conservationmagazine.org/2008/07 /a-good-fish-for-the-wine/print/.

Merchant, Carolyn. *The Death of Nature: Women, Ecology, and the Scientific Revolution*. San Francisco: Harper Row, 1993.

Merleau-Ponty, Maurice. *Phenomenology of Perception*. London: Routledge and Kegan Paul, 1962.

Michaels, Anne. *The Winter Vault*. New York: Alfred Knopf, 2009.

Migaud, Herve, John Taylor, and Tom Hansen. "Sterile Salmon: Toward a More Sustainable and Eco-friendly Industry." 2008. Accessed April 25, 2012, http://www.salmotrip.stir.ac.uk/downloads/FishFarmer.pdf.

Miller, Arthur. "The Year It Came Apart." *New York Magazine*, vol. 8, no. 1, December 30, 1974.

Mills, Stephanie. *Tough Little Beauties: Selected Essays and Other Writings*. North Liberty, Iowa: Ice Cube Press, 2007.

———. *Turning Away from Technology: A New Vision for the 21st Century*. San Francisco: Sierra Club Books, 1997.

Monbiot, George. "Why Whale Poo Matters." *Guardian*, December 12, 2014. https://www.theguardian.com/environment/georgemonbiot/2014/dec/12/how-whale-poo-is-connected-to-climate-and-our-lives.

Montgomery, David. *Dirt: The Erosion of Civilizations*. Berkeley, CA: University of California Press, 2007.

———. *King of Fish: The Thousand Year Run of the Salmon*. Cambridge, MA: Westview Press, 2003.

Mueller, Martin Lee. *Symphony of Silences*. Saarbruecken: VDM, 2009.

Muir, John. *John of the Mountains: The Unpublished Journals of John Muir*. Edited by Linnie Marsh Wolfe. Boston: Houghton Mifflin, 1938.

Mumford, Lewis. *The Pentagon of Power*. New York: Hartcourt, Brace, 1970.

Murray, Rupert (director). *The End of the Line. Imagine a World Without Fish*. Arcane Pictures, Calm Productions, and Dartmouth Films, 2009. Accessed June 1, 2014, http://endoftheline.com/.

Nadler, Steven. "Baruch Spinoza." *Stanford Encyclopedia of Philosophy*. Accessed January 23, 2013, http://plato.stanford.edu.

———. *Spinoza: A Life*. Cambridge, UK: Cambridge University Press, 2001.

Næss, Arne. *Deep Ecology of Wisdom/Selected Works of Arne Næss*. Edited by Harold Glasser and Alan Drengson. Dordrecht: Springer, 2005.

———. *Ecology, Community and Lifestyle*. Cambridge, UK: Cambridge University Press, 1989.

———. "Self-Realization: An Ecological Approach to Being in the World." In *The Selected Works of Arne Næss*, edited by Harold Glasser and Alan Drengson. Dordrecht: Springer, 2005.

Nagel, Thomas. "What Is It Like to Be a Bat?" *The Philosophical Review*, vol. 83, no. 4 (1974): 435–450. doi:10.2307/2183914.

Nelson, Richard. *The Island Within*. New York: Vintage Books, 1991.

Noble, David. *The Religion of Technology: The Divinity of Man and the Spirit of Invention*. New York: Alfred Knopf, 1998.

Northcott, Michael S. *A Political Theology of Climate Change*. Grand Rapids, MI: Wm. B. Eerdmans Publishing Co., 2013.

Oliver, Mary. "Sleeping in the Forest." *Ohio Review*, vol. 19, no. 1 (winter 1978): 49–60.

Orr, David. *Earth in Mind: On Education, Environment, and the Human Prospect*. Washington, DC: Island Press, 2004.

Osawa, Sandra. "The Politics of Taking Fish." In *First Fish, First People Salmon Tales of the North Pacific Rim*, edited by Judith Roche and Meg McHutchison. Seattle: University of Washington Press, 1998.

Ou, Michelle, Trevor J. Hamilton, Junho Eom, Emily M. Lyall, Joshua Gallup, Amy Jiang, Jason Lee, David A. Close, Sang-Seon Yun, and Colin J. Brauner. "Responses of Pink Salmon to CO_2-Induced Aquatic Acidification." *Nature Climate Change*, vol. 5 (October 2015): 950. doi:10.1038/nclimate2694.

The Oxford English Dictionary. 2nd edition. Oxford: Clarendon Press, 1989.

The Oxford Universal Dictionary. London: Clarendon Press, 1973.

Øyehaug, Ogne. "Vosso-laks i Trønderasyl." *Bergens Tidene*, 2007. Accessed April 15, 2012, http://www.bt.no/na24/article350726.ece.

PETA. "Commercial Fishing: How Fish Get from the High Seas to Your Supermarket." Accessed November 17, 2013, http://www.peta.org/issues /animals-used-for-food/commercial-fishing.aspx.

Peterson, Anna L. *Being Human: Ethics, Environment, and Our Place in the World*. Berkeley, CA: University of California Press, 2003.

Pollan, Michael. *The Botany of Desire: A Plant's Eye View of the World*. New York: Random House, 2001.

Poulter, Sean. "Scientists Create GM 'Frankenfish' Which Grows Three Times as Fast as Normal Salmon." *Daily Mail*, June 17, 2010. http://www.dailymail.co .uk/news/article-1287084/Scientists-create-GM-Frankenfish-grows-times -fast-normal-salmon.html.

Reiter, Reinhard. "Die Lachsforelle: Großwüchsig, Rotfleischig und Reich an Hochwertigen Fetten." Accessed November 1, 2014, http://www.lfl.bayern.de.

Rifkin, Jeremy. "The Empathic Civilization: An Address Before the British Royal Society for the Arts." March 15, 2010. Retrieved from Bichara Sahely's blog, May 9, 2017, https://bsahely.com/2012/06/24/the_empathic_civilization/.

———. *The Empathic Civilization: The Race to Global Consciousness in a World in Crisis*. New York: Tarcher Penguin, 2010.

———. *The European Dream: How Europe's Vision of the Future Is Quietly Eclipsing the American Dream*. Cambridge, UK: Polity Press, 2004.

————. "RSA Animate: The Empathic Civilization." YouTube video, 10:39. Posted May 6, 2010. http://www.youtube.com/watch?v=l7AWnfFRc7g.

Ripple, William J., Thomas P. Rooney, and Robert L. Beschta. "Large Predators, Deer, and Trophic Cascades in Boreal and Temperate Ecosystems." In *Trophic Cascades*, edited by John Terborgh and James A. Estes. Washington, DC: Island Press, 2009.

Rockström, Johan, Will Steffen, Kevin Noone, Åsa Persson, F. Stuart III Chapin, Eric Lambin, Timothy M. Lenton, et al. "Planetary Boundaries: Exploring the Safe Operating Space for Humanity." *Ecology and Society*, vol. 14, no. 2 (2009): 32. http://www.ecologyandsociety.org/vol14/iss2/art32/.

Roman, Joe, and James J. McCarthy. "The Whale Pump: Marine Mammals Enhance Primary Productivity in a Coastal Basin." *PLoS ONE*, vol. 5, no. 10 (2010): e13255. doi:10.1371/journal.pone.0013255.

The Royal Society of Canada. "Elements of Precaution: Recommendations for the Regulation of Food Biotechnology in Canada, Ottawa." January 2001. Accessed March 4, 2012, http://www.rsc.ca/en/expert-panels/rsc-reports /elements-precaution-recommendations-for-regulation-food-biotechnology-in.

Rozell, Ned. "Salmon Nose Deep into Alaska Ecosystems." Accessed November 1, 2010, http://www.gi.alaska.edu/ScienceForum/ASF17/1721.html.

Rubenstein, Mary-Jane. *Strange Wonder: The Closure of Metaphysics and the Opening of Awe*. New York: Columbia University Press, 2008.

Ruby, Robert H., John A. Brown, and Cary C. Collins. *A Guide to the Indian Tribes of the Pacific Northwest*. Norman: University of Oklahoma Press, 2010.

Rust, Mary-Jayne, and N. Totton, eds. *Vital Signs: Psychological Responses to Ecological Crisis*. London: Karnac Books, 2011.

Savoca, Matthew S., and Gabrielle A. Nevitt. "Evidence That Dimethyl Sulfide Facilitates a Tritrophic Mutualism between Marine Primary Producers and Top Predators." *Proceedings of the National Academy of Sciences of the United States of America* (2013). doi:10.1073/pnas.1317120111.

Schätzing, Frank. *Nachrichten aus Einem Unbekannten Universum. Eine Zeitreise Durch die Meere*. Cologne: Kiepenheuer and Witsch, 2006.

Schiller, Friedrich von. "Was Heißt und Zu Welchem Ende Studiert Man Univer- salgeschichte?" 1789. Accessed December 1, 2014, http://www.ub.uni-bielefeld .de/diglib/aufkl/teutmerk/teutmerk.htm.

Schmitz, Oswald J., Peter A. Raymond, James A. Estes, Werner A. Kurz, Gordon W. Holtgrieve, Mark E. Ritchie, Daniel E. Schindler, et al. "Animating the Carbon Cycle." *Ecosystems*, vol. 17, no. 2 (2014): 344–359. doi:10.1007 /s10021-013-9715-7.

SeaWeb. "Bottom Trawling Impacts, Clearly Visible from Space." February 15, 2008. Accessed November 7, 2011, http://www.eurekalert.org/pub_releases /2008-02/s-bti021508.php.

Shakespeare, William. *Family Shakespeare in Ten Volumes.* 2nd edition. Edited by Thomas Bowdler. London: Longman, Hurst, Rees, Orme and Brown, 1820.

Shelley, Mary W. *Frankenstein, or: The Modern Prometheus.* Philadelphia: Courage Books, 1818/1990.

Shepard, Paul. *Encounters with Nature: Essays.* Washington, DC: Island Press, 1999.

———. *Nature and Madness.* Athens, GA: University of Georgia Press, 1998.

Shubin, Neil. *Your Inner Fish.* New York: Random House, 2009.

Simard, Suzanne. *Mother Tree.* October 8, 2012. Video footage. http://www .ecology.com/2012/10/08/trees-communicate/

Sloterdijk, Peter. *Sphären,* 3 vols. Frankfurt am Main: Suhrkamp, 1998–2004.

Smith, Adeline. *Boston Charlie and Bigfoot.* Compiled and edited by Jamie R. Valadez. Lower Elwha Klallam Tribe, 2011.

Snow, Charles Percy. *The Two Cultures.* London: Cambridge University Press, 1959/1998.

Soechtig, Stephanie. *Fed Up.* Documentary film. Atlas Films, January 19, 2014. http://fedupmovie.com/.

Temple, William. *Nature, Man and God.* Edinburg: R and R, Clark, Ltd., 1934.

Teilhard de Chardin, Pierre. *The Heart of Matter.* San Diego: Harcourt, 1978.

Thompson, Lucy (Che-ne-wah Weitch-ah-wah). *To the American Indian: Reminiscences of a Yurok Woman.* Berkeley, CA: Heyday Books, 1916.

Thoreau, Henry David. *Walden.* Princeton, NJ: Princeton University Press, 1971.

Thorstad, Eva, Ian A. Fleming, Philip McGinnity, Doris Soto, Vidar Wennevik, and Fred Whoriskey. "Incidence and Impacts of Escaped Farmed Atlantic Salmon *Salmo salar* in Nature." NINA special report 36 (2008). Retrieved from FAO October 1, 2012, ftp://ftp.fao.org/fi/document/aquaculture/aj272e00.pdf.

Todal, Per Anders. "Spår Eksplosiv Oppdrettsvekst." *Dag og Tid*, November 16, 2012. http://www.dagogtid.no/nyhet.cfm?nyhetid=2415.

Tucker, Abigail. "On the Elwha, a New Life When the Dam Breaks." *Smithsonian*, September 14, 2012. http://www.smithsonianmag.com/people-places/On-the -Elwha-a-New-Life-When-the-Dam-Breaks.html?c=y&page=1.

———. "Preparing for a New River." *Smithsonian*, December 2011. http://www .smithsonianmag.com/people-places/preparing-for-a-new-river-104398/.

US General Accounting Office. "Developing Markets for Fish Not Traditionally Harvested by the United States: The Problems and the Federal Role." May 7, 1980. CED-80-73.

Valadez, Jamie. *Boston Charlie and Bigfoot: Told by Adeline Smith.* Lower Elwha Klallam Reservation: Lower Elwha Klallam Tribe, 2011.

———. *A Collection of Klallam Stories.* Port Angeles, WA: Lower Elwha Klallam Tribe, 2008.

———. *The Elwha River and Its People.* Port Angeles, WA: Lower Elwha Klallam Tribe, 2008.

————. *Klallam Stories, told by Adeline Smith*. Lower Elwha Klallam Reservation: Lower Elwha Klallam Tribe, 2012.

————. *Naturally Native: Klallam Gathering, Hunting and Fishing*. Lower Elwha Klallam Reservation: Lower Elwha Klallam Tribe, 2011.

Valadez, Jamie, and Carmen-Watson Charles. *Tse-whit-zen: A Klallam Village*. Port Angeles, WA: Lower Elwha Klallam Tribe, 2008.

Vetlesen, Arne Johan. "Ingen Vei Ut?" *Klassekampen*, June 12, 2012.

————, ed. *Nytt Klima. Miljøkrisen i Samfunnskritisk Lys*. Oslo: Gyldendal Norsk Forlag, 2008.

Vidal, John. "Protect Nature for World Economic Security, Warns UN Biodiversity Chief." *Guardian*, August 16, 2010. http://www.theguardian.com /environment/2010/aug/16/nature-economic-security.

Walbran, John T. *British Columbia Coast Names, 1592–1906: Their Origin and Ancestry*. Vancouver: J.J. Douglas Lt., 1909/1971.

Waldau, Paul, and Kimberly Patton, eds. *A Communion of Subjects: Animals in Religion, Science and Ethics*. New York: Columbia University Press, 2006.

Walrond, Carl. "Trout and Salmon—Chinook Salmon." *Te Ara—the Encyclopedia of New Zealand*. November 24, 2008. Accessed May 10, 2017, http://www .TeAra.govt.nz/en/trout-and-salmon/page-3.

Walton, Izaak. *The Compleat Angler*. London: Studio Editions, 1653/1990.

Warren, Brad. *Ocean Acidification: Richard Feely Shows How It Might Kill the Fishing Business*. 2007. Accessed December 2, 2011, http://www.pacificfishing .com/climate_change/ocean_acidification.pdf.

Waters, Colin N., Jan Zalasiewicz, Colin Summerhayes, Anthony D. Barnosky, Clément Poirier, Agnieszka Gałuszka, Alejandro Cearreta, et al. "The Anthropocene Is Functionally and Stratigraphically Distinct from the Holocene." *Science*, vol. 351, no. 6269 (January 8, 2016). doi:10.1126/science.aad2622.

Webb, Caroline. "Thomas Berry on Nature and Humans." YouTube video, 09:28. Posted October 2008.

Weber, Andreas. *Matter and Desire: An Erotic Ecology*. White River Junction, VT: Chelsea Green Publishing, 2017.

Wertsch, James V. *Voices of Collective Remembering*. Cambridge, UK: Cambridge University Press, 2002.

Weston, Anthony. "Multicentrism: A Manifesto." *Environmental Ethics*, vol. 26, no. 1 (2004): 25–40.

White, Richard. *The Organic Machine: The Remaking of the Columbia River*. New York: Hill and Wang, 1996.

Whitman, Walt. *Leaves of Grass*. New York: New York University, 1965.

Wolf, Edward C., and Seth Zuckerman, eds. *Salmon Nation: People, Fish, and Our Common Home*. Portland, OR: Ecotrust, 2003.

Wood, David. "Who Do We Think We Are? Wonder in the Anthropocene."
Edited by Joe Gelonesi. ABC, April 24, 2015. Accessed August 1, 2015,
http://www.abc.net.au/radionational/programs/philosopherszone/who-do-we
-think-we-are/6415066.

Wray, Jacilee, ed. *Native Peoples of the Olympic Peninsula: Who We Are*. Norman:
University of Oklahoma Press, 2003.

Ye, Yimin, and Kevern Cochrane. "Global Overview of Marine Fishery
Resources." *FAO Fisheries and Aquaculture Technical Paper*, 569 (2011): 6.
Accessed May 16, 2017, http://www.fao.org/docrep/015/i2389e/i2389e.pdf.

Ystad, Vidar. "Fiskerisjefar Tek Laksen På Land." *Dag og Tid*, July 7, 2011. http://
old.dagogtid.no/nyhet.cfm?nyhetid=2060.

Zimov, S. A., E. G. Schuur, and F. S. Chapin III. "Permafrost and the Global
Carbon Budget." *Science*, vol. 312, no. 5780 (2006): 1612–1613.

Zinn, Howard. *A People's History of the United States*. New York: Harper
Perennial, 2010.

NOTES

Preface

1. Among many others, vivid examples include the works of theologian Thomas Berry, geophilosopher and ecologist David Abram, Buddhist scholar and philosopher Joanna Macy, veterinarian Bergljot Børresen, social philosopher Charles Eisenstein, the writer Neil Evernden, the educator David Orr, the Buddhist monk Thich Nhat Hanh, Earth lawyer Polly Higgins, the ecopsychologist David Fisher, and the pioneering ecophilosopher Arne Næss.
2. Wood, "Who Do We Think We Are?"
3. Thank you, Ernesto Adrian Lara, for pointing this out. The full quote from Aristotle's *Metaphysics* goes as follows: "That [philosophy] is not a productive science is clear from a consideration of the first philosophers. It is through wonder that men now begin and originally began to philosophize; wondering in the first place at obvious perplexities, and then by gradual progression raising questions about the greater matters too, e.g. about the changes of the moon and of the sun, about the stars and about the origin of the universe. Now he who wonders and is perplexed feels that he is ignorant (thus the myth-lover is in a sense a philosopher, since myths are composed of wonders)." Aristotle, *Metaphysics*, 1.982b.
4. Wood, "Who Do We Think We Are?"
5. Ibid.
6. Ibid.
7. See Harding and Margulis, "Water Gaia." I will relate this story in some detail in chapter 7.
8. "Cycles of participation" is an expression coined by the ecologist Stephan Harding, in lieu of the more commonly known, rather more technical expression "feedback loops."
9. "Depth" is a phenomenological term which Abram has explored at some length.
10. Quoted in Callicott, *Earth's Insights*, 183.
11. Evernden, *Natural Alien*, 141.
12. Hannesson, "Hva Skal Vi med Villaksen?" (author's translation).
13. Aldwell, *Conquering the Last Frontier*, 80.
14. Ibid., 14.

15. William Ware, *Seattle Post-Intelligencer,* December 1, 1901, quoted in Mapes, *Elwha: A River Reborn,* 53.
16. Walton, *The Compleat Angler,* 148.
17. Johansson, *Laks,* 37. In the original, it reads: "at laksen holdes som seg bør for den edleste og den beste og vakreste som fanges i Norge" (author's translation).

Chapter 1

1. Temple, *Nature, Man and God,* 57.
2. Descartes, *Discourse on Method.*
3. Ibid.
4. Ibid.
5. Aquinas, *Quaestiones Disptatae de Veritate.*
6. Drengson, *Wild Way Home,* 170.
7. In August 2016, the Working Group on the Anthropocene (WGA) declared that the accumulated human impact on the biosphere has taken dimensions of such profound consequences that the Holocene stable state has in fact ended, giving way to a new Earth epoch, the Anthropocene or "the Age of Humans" (a name popularized by Nobel laureate Paul Crutzen).

    Thomas Berry, in conversation with Brian Swimme, once coined an alternative name for this new Earth epoch: the Ecozoic! Combining *eco* (Greek *oikos*: "house," "household," or "home") and *zoic* (Greek *zoikos*: "pertaining to living beings"), the Ecozoic would be defined as "the era of the house of living beings" (Kiplinger, "Groping Our Way Towards"), or the era in which *all* life is making itself at home again inside Earth. The name would have marked a clear cut with the very long anthropocentric tradition, prompting richer conversations about how to reintegrate the human into the vastly more-than-human biotic community.
8. Wertsch, *Voices of Collective Remembering,* 56.
9. Ibid.
10. Berry, *Dream of the Earth,* xi.
11. See Glendinning, *My Name is Chellis,* and Harding, *Animate Earth,* 31f.
12. The following paragraphs are inspired by David Abram, who drew these connections during the University of Oslo's spring course *Philosophical Foundations of Ecomodernity* in 2014. See Abram, "In the Depths of a Breathing Planet."
13. See Wood, "Who Do We Think We Are?"
14. My gratitude to David Abram, who prompted me to think carefully about what implications that profound distrust of the senses would have on the era of early modern thought (personal conversation).
15. Quoted in Guignon, *Heidegger and the Problem of Knowledge,* 23.

16. I am wary of my own metaphor. It might suggest that human existence prior to that shattering moment simply could be reduced to a kind of gestation, not quite as real as what happened afterward, which of course it cannot and should not.

Chapter 2

1. The presentation was originally made in Norwegian. Its title was *Jakten på Gladlaksen*. All citations are my translation.
2. I have yet to find a formula that lets me calculate how many women it would take to fill up Rica Hell Hotel. All the calculations available on the internet at the time of my search were tailored to men.
3. See Egeland, "Nå Dreper de Fjordene."
4. Guignon, *Heidegger and the Problem of Knowledge*, 17 (emphasis added).
5. Matthews, "Letting the World Grow Old," 227.
6. Ibid.
7. See Merchant, *Death of Nature*; Spretnak quoted in Mills, *Tough Little Beauties*, 163ff.
8. Northcott, *Political Theology of Climate Change*, 190.
9. Latour first made this somewhat provocative point in his small monograph *We Have Never Been Modern*.
10. See Martinsen, "Fisken, det Usynlige Dyret."
11. Geelmuyden, *Sannheten på Bordet*, 79–80 (author's translation).
12. Krogshus, "Norsk Laks: Det Dødelige Pengeeventyret."
13. The name commonly associated with the true breakthrough of environmental thought into the public sphere is Rachel Carson. Her hugely influential book *Silent Spring* (1962) has been credited not only for inspiring the grassroots, activist environmental movement, but also for giving rise to ecophilosophy as a distinct discipline, with various schools that have since emerged from within. In the final passage of her book, Carson wrote: "The 'control of nature' is a phrase conceived in arrogance, born of the Neanderthal age of biology and philosophy, when it was supposed that nature exists for the convenience of man." *Silent Spring*, 297.
15. Guignon, *Heidegger and the Problem of Knowledge*, 70.
16. Guignon says to this point: "Like the interpretation of a text, the interpretation of Dasein must always be circular. There are no axioms or self-evident truths from which we can build up an edifice of knowledge about ourselves. As our lives always invoke a back-and-forth movement between partial meanings and some sense of the whole, the method of fundamental ontology also moves back and forth between uncovering structural items . . . and a pre-understanding of the totality. . . . The hermeneutic circle is constitutive of Dasein's Being." Guignon, *Heidegger and the Problem of Knowledge*, 71. See Bruce Foltz, *Inhabiting Earth*.

17. Foltz, *Inhabiting Earth*, 103.

18. Guignon, *Heidegger and the Problem of Knowledge*, 16.

19. Quoted in Guignon, *Heidegger and the Problem of Knowledge*, 16.

20. In this sense, what I have called "human exceptionalism" will also show itself to be a particular version of narrative exceptionalism, or hegemony. The unifying element of the dualisms "human"/"other-than-human," "mind/body," and "modern"/"other-than-modern" is the ambition to "otherize" that which is perceived to be outside.

21. Magnussøn, *Making Food*, 67.

22. Quoted in Poulter, "Scientists Create GM 'Frankenfish.'"

23. AquaBounty Technologies, Inc., "Environmental Assessment for AquAdvantage® Salmon," (emphasis added).

24. Kearns and Mayers, "Substantial Equivalence."

25. Brunner et al., "Beyond 'Substantial Equivalence.'"

26. Royal Society of Canada, "Elements of Precaution," 182.

27. After having served the FDA a second time, and after having fortified the principle of substantial equivalence in FDA policies, Taylor returned once again to Monsanto, this time to become its vice president. But even that was not his last walk through the infamous revolving door. In July 2009, he joined the FDA a third time, this time under the Obama administration, serving first as senior advisor to the FDA commissioner, and then, half a year later, assuming another newly created post, that of deputy commissioner for foods. This third tenure of Taylor's with the FDA coincided with the passing of the so-called S510 Food Safety Modernization Act. The act grants the FDA unprecedented power over the entire food production process: "No Limit on Secretarial Authority—Nothing in this section shall be construed to limit the ability of the Secretary to review and act upon information from food testing, including determining the sufficiency of such information and testing." (S510 (111th): FDA Food Safety Modernization Act, http://www.govtrack.us /congress/bills/111/s510/text) In other words, the FDA decides if, when, where, and what types of food tests will be undertaken. It decides also what test results will be made available, "in the interest of national security." It would seem that the best way to act in the interest of national security would be to publish *all* tests, and then let the public decide for themselves how they wish to handle the results. Whose security—whose interest—are you serving when you have absolute power to withhold certain test results, and to publish others?

28. Alliance for Natural Health, "Action Alert: Genetically Modified Frankenfish!" (emphasis added).

29. AquaBounty Technologies, Inc., "AquaBounty Fact Sheet."

30. AquaBounty Technologies, Inc., "Substantial Equivalence," 10.

31. AquaBounty Technologies, Inc., "AquaAdvantage Fish."
32. See Soechtig, *Fed Up*.

Chapter 3

1. The authors also note that this is a rather novel arrangement. Earlier forms of domestication typically favored animals that were *less* hungry than their undomesticated heirs, for the hungrier they were, the greater a strain they would be on their keepers.
2. Lien and Law, "Emergent Aliens," 8.
3. Ibid., 16.
4. Ibid., 4.
5. Ibid., 6.
6. Ibid., 17.
7. Clutton-Brock, *Natural History*, 21.
8. Lien and Law, "Emergent Aliens," 10.
9. Ibid., 12.
10. Ibid.
11. In fact, to this day the legal definition of the word *domicile*—a cousin of *domestication* that is derived from the same Latin root, *domus*—emphasizes just that aspect of permanence. "Domicile," according to the *Oxford Universal Dictionary*, is defined in law as "the place where one has his permanent residence to which, when absent, he has the intention of returning."
12. *Oxford Universal Dictionary*.
13. Keith, *Vegetarian Myth*, 25–26.
14. Ibid.
15. Pollan, *Botany of Desire*, xvi; xvi–xvii.
16. Ibid.
17. Ibid.
18. Undomesticated salmon practically never eat a vegetarian diet.
19. As we will explore at more depth in later chapters, the Klallam and other indigenous cultures think of the salmon's return to their birth rivers as an act of *choice*, precisely as a *gift* that the salmon make to the more-than-human riparian community, including to human riverside dwellers. This gift-giving aspect holds central significance in these cultures' understanding of salmon agency. In contrast, the salmon industry is designed to be intrinsically ignorant of this aspect of salmon agency.
20. Magnussøn. *Making Food*, 64 (emphasis added).
21. Ibid., 67, 53.
22. Lien and Law, "Emergent Aliens," 10.

23. Magnussøn, *Making Food*, 66.

24. Ibid., 73–74.

25. Ibid., 74–75.

26. See Greenberg, *Four Fish*.

27. See Øyehaug, "Vosso-laks i Trønderasyl."

28. If we wished to study power relations in this specific case, it would not be so meaningful to speak of the absence or presence of reciprocity. It would seem more relevant to discuss that particular case as an instance of *empowerment*. Empowerment, too, is a form of power.

29. Migaud et al., "Sterile Salmon."

30. Ibid.

31. Ibid.

32. For their project SALMOTRIP, the researchers secured funding by these industry partners: Landcatch Natural Selection, Aquagen, the Centre for Aquaculture Competence, Salmo, and Marine Harvest.

33. Foster, "Salmon Successfully Farmed Inland."

34. Ibid.

35. Quoted in Langberg, "Kina Skal Produsere Laks i Ørkenen."

36. Berg-Hansen, "Laks på Land" (author's translation).

37. Quoted in Ystad, "Fiskerisjefar Tek Laksen på Land," (author's translation and emphasis added).

38. Or at least most of them will; the technology is only considered 99 percent effective.

39. AquaBounty Technologies Inc., "Environmental Assessment."

40. Ibid.

41. Thorstad et al., "Incidence and Impacts," 3 (emphasis added).

42. Ibid., 6.

43. See Greenberg, *Four Fish*.

44. They also changed in a number of other ways, ways that were not explicitly chosen for but that developed rather unintentionally alongside the ones already mentioned. Among these are deformities, different spawning behavior and success, as well as changes in morphology, spawning time, fertility, and egg viability. See Thorstad et al., "Incidence and Impacts."

45. Thorstad et al., "Incidence and Impacts," 6.

46. Børresen, *Fiskenes Ukjente Liv*, 39 (author's translation).

47. Ibid.

48. But this created a new kind of dilemma for these caretakers. For eventually people *did* begin to kill their domesticated animal companions for food, all the while continuing to forge strong emotional bonds with those same companions while they lived around humans. It was a dilemma that had no simple solutions. But by and large, Børresen shows, humans appear to have

found ways to respectfully kill individual animals for food without seriously diminishing their emotionally charged, caring, and empathetic social bonding with domesticated animals.

49. Clutton-Brock, *Walking Larder*, 22 (emphasis added).
50. Børresen, *Fiskenes Ukjente Liv*, 43.
51. House, *Totem Salmon*, 99.
52. It also shows that the Cartesian split of the thinking mind from the body is being mobilized by the industry *only* to the degree that it serves the industry's purposes. And vice versa: Emotive ways of knowing—feelings—are also called upon to the degree that those serve the industry's purposes.
53. Børresen, *Fiskenes Ukjente Liv*, 57.
54. Ibid. Børresen gives the following example: There are animals whose facial expressions and body language we can easily interpret. When those animals scream, writhe, and kick in pain, it is (typically) obvious to us that these beings are suffering. This is not as easy with fish. There is a strong incentive among fish, writes Børresen, to *not* give any signs of weakness even during the most excruciating of pains. For there might always be a larger predator in the waters who would pick up the signs and pose a potential threat. Not showing pain, for fish, is a way to protect themselves. (See Børresen, *Fiskenes Ukjente Liv*, 72f.)
55. See Børresen, *Fiskenes Ukjente Liv*.
56. See *Oxford English Dictionary*.
57. The category "tameness" does not apply because salmon, if given the choice, will consistently escape their enclosures.
58. The invocation of the trope "farm"—common though it is in the industry—is rather problematic, given the trope's strong connotations with deeply intimate, personal relationships between the farmers and their animals.
59. Børresen's suggestion to think of all industrially produced animals, including Atlantic salmon, as "exploited captives" takes on an uncanny and disturbing timbre when we consider the sheer number of animals who are actually being kept under conditions of industrial mass production. This is the *Guardian* writer Yuval Noah Harari: "In 2009, there were 1.6 billion wild birds in Europe, counting all species together. That same year, the European meat and egg industry raised 1.9 billion chickens. Altogether, the domesticated animals of the world weigh about 700m tonnes, compared with 300m tonnes for humans, and fewer than 100m tonnes for large wild animals." Harari follows this up with a remarkable and lucid argument: "[T]he fate of farm animals is not an ethical side issue. *It concerns the majority of Earth's large creatures*: tens of billions of sentient beings, each with a complex world of sensations and emotions, but which live and die on an industrial production line." Harari, "Industrial Farming" (emphasis added).

60. Eisenstein, *More Beautiful World*, 213.
61. Ibid.

Chapter 4

1. Callicott, *In Defense of the Land Ethic*, 102.
2. Ibid., 101.
3. Callicott, *Earth's Insight*, 84.
4. Callicott, *In Defense of the Land Ethic*, 104.
5. Shepard, *Encounters with Nature*, 69.
6. Quoted in Callicott, *In Defense of the Land Ethic*, 105.
7. Hannesson, "Hva skal vi med villaksen?" (author's translation).
8. Ibid.
9. Rozell, "Salmon Nose Deep into Alaska Ecosystems."
10. Meadows, "A Good Fish for the Wine."
11. See Montgomery, *King of Fish*.
12. See Helfield and Naiman, "Effects of Salmon-Derived Nitrogen."
13. Quoted in Callicott, *In Defense of the Land Ethic*, 105.
14. Callicott, *In Defense of the Land Ethic*, 105–106.
15. Shepard, *Encounters with Nature*, 69.
16. Quoted in Callicott, *In Defense of the Land Ethic*, 107.
17. Shepard, *Encounters with Nature*, 69; see also Callicott, *In Defense of the Land Ethic*, 108.
18. Quoted in Callicott, *In Defense of the Land Ethic*, 108.
19. See Harding, *Animate Earth*.
20. See Christ, "Intimations of Gaia."
21. Personal conversation.
22. Næss, *Deep Ecology of Wisdom*, 7.
23. Berry, *Story of the Universe*, 77–78.
24. Jensen, *Thought to Exist in the Wild*, 86–87.
25. See Doughty et al., "Global Nutrient Transport in a World of Giants." According to a report by Ted Gresh, Jim Lichatowich, and Peter Schoonmaker, Pacific salmon brought an annual amount of 160 million to 226 million kilograms of fertilizers to the Pacific Northwest before their numbers began to collapse. By the year 2000, these numbers had dwindled to between 11.8 and 13.7 million kilograms, or roughly 6 to 7 percent of what once reached the rivers of the Pacific Northwest. Gresh et al., "Salmon Decline Creates Nutrient Deficit."
26. Anadromous fish are those who are born in fresh water, spend their adult life in the oceans and return again to fresh water to spawn (from Greek: ἀνά, *ana*,

"up" and δρόμος, *dromos*, "course"). Next to salmon, other familiar anadromous fish include shad, striped bass, smelt, or sturgeon.

27. Doughty et al., "Global Nutrient Transport in a World of Giants."
28. Ibid.
29. Ed Hunt, contributor to the community network Salmon Nation, has gathered a list of as many as 137 animal species from along the Pacific Rim who all rely on Pacific salmon. Here are the other animals he lists:

osprey, Caspian tern, black bear, northern river otter, killer whale, Cope's giant salamander, red-throated loon, pied-billed grebe, Clark's grebe, American white pelican, Brandt's cormorant, double-crested cormorant, pelagic cormorant, great blue heron, black-crowned night-heron, turkey vulture, California condor, common goldeneye, Barrow's goldeneye, common merganser, red-breasted merganser, golden eagle, Bonaparte's gull, Heermann's gull, ring-billed gull, California gull, herring gull, Thayer's gull, western gull, glaucous-winged gull, glaucous gull, common tern, arctic tern, Forster's tern, elegant tern, common murre, marbled murrelet, rhinoceros auklet, tufted puffin, belted kingfisher, American dipper, Steller's jay, black-billed magpie, American crow, northwestern crow, common raven, Virginia opossum, water shrew, raccoon, mink, bobcat, northern fur seal, northern (steller) sea lion, California sea lion, harbor seal, Pacific white-sided dolphin, gyrfalcon, peregrine falcon, killdeer, spotted sandpiper, snowy owl, willow flycatcher, tree swallow, violet-green swallow, northern rough-winged swallow, bank swallow, cliff swallow, barn swallow, harbor porpoise, Dall's porpoise, snapping turtle, western pond turtle, western terrestrial garter snake, common garter snake, Pacific loon, common loon, yellow-billed loon, horned grebe, red-necked grebe, western grebe, sooty shearwater, brown pelican, great egret, snowy egret, green heron, trumpeter swan, mallard, green-winged teal, canvasback, greater scaup, surf scoter, white-winged scoter, hooded merganser, red-tailed hawk, greater yellowlegs, Franklin's gull, mew gull, black-legged kittiwake, pigeon guillemot, ancient murrelet, gray jay, winter wren, American robin, varied thrush, spotted towhee, song sparrow, masked shrew, vagrant shrew, montane shrew, fog shrew, Pacific shrew, Pacific water shrew, Trowbridge's shrew, Douglas' squirrel, deer mouse, red fox, gray fox, ringtail, American marten, fisher, long-tailed weasel, wolverine, striped skunk, mountain lion, white-tailed deer, black-tailed deer, minke whale, sperm whale, humpback whale, and northern rightwhale dolphin. Hunt, "137 Species Rely on Pacific Salmon."

30. Cederholm et al., "Fate of Coho Salmon."
31. Gresh et al., "Salmon Decline."
32. Ibid.
33. Ibid.

34. Ibid.

35. See Bump et al., "Wolves Modulate Soil Nutrient Heterogeneity."

36. Ripple and his colleagues write: "Crête and Manseau (1996) found moose (*Alces alces*) densities in eastern North America to be seven times higher in a region without wolves (1.9 moose/km2) compared to one with wolves (0.27 moose/km2), even though primary productivity was higher in the area with wolves." Ripple et al., "Large Predators, Deer, and Trophic," 3.

37. Schmitz and his colleagues draw an interesting conclusion from their findings. They suggest that letting the wolves assume their place as keystone species of the forests again could help to restore the forests as carbon sink: "Part of this . . . effort could include ensuring that apex predators such as wolves attain natural levels that maintain moose populations below their carrying capacity." Schmitz et al., "Animating the Carbon Cycle."

38. Stephan Harding has suggested that the final outcome of wolf predation (warming or cooling?) might be so complex that we will never be able to quantify it. There are likely to be great variations with regard to time and space, and a great number of variables, including population dynamics of wolves, herbivores, and the different species of trees. Personal conversation.

39. There are organisms who can synthesize food without sunlight. Typically these organisms use some form of chemosynthesis, or the production of original nutrients from high-energy, inorganic compounds such as hydrogen gas, ammonia, nitrates, or sulfides. Among the known sources of such inorganic compounds are deep-sea hydrothermal vents, small cracks in the Earth's mantle far beyond the reach of sunlight, in the eternal darkness of the deep oceans. Such vents are an example for a habitat that can thrive fully out of reach of the sun, nourished only by the Earth's molten core.

40. See Roman and McCarthy, "The Whale Pump."

41. See Lavery et al., "Iron Defecation by Sperm Whales"; and Monbiot, "Why Whale Poo Matters." The discussion offers a way in which an oceanwide rebounding of marine mammals becomes an imperative even from a strictly anthropocentric perspective. One needs not *necessarily* invoke a more inclusive position—be it biocentric, ecocentric, holistic, or multicentric—to make a strong case for a comprehensive return of marine mammals to pre-exploitation abundance.

42. See Savoca and Nevitt, "Evidence That Dimethyl Sulfide Facilitates"; and Monbiot "Why Whale Poo Matters."

43. See Charlson et al., "Oceanic phytoplankton."

44. Evernden, *Natural Alien*, 133.

45. As I write this, Norway is witnessing a furious debate around its wolves. Of the country's total wolf population of sixty-eight animals, the government is

suggesting to kill forty-seven this winter. Gunnar Gunerdsen, a parliamentarian for the right-wing party, has said that "this case is not about the law. It's all about politics," echoing prime minister Erna Solberg, who is among those who have suggested that both Norway's existing national law on biological diversity and the Bern Convention ought to be simply ignored.

Chapter 5

1. For the evolution of breasts, teeth, bones, hair, and feathers from skin, read "Teeth Everywhere" in Shubin, *Your Inner Fish*.
2. Margulis and Sagan, *Microcosmos*, 183–184.
3. Ibid.
4. Abram, *Spell of the Sensuous*, ix.
5. Ibid., 22.
6. Nadler, "Baruch Spinoza."
7. Abram, *Becoming Animal*, 107.
8. Ibid., 108.
9. Nadler, *Spinoza: A Life*, 120.
10. Quoted in Duquette, *Hegel's History of Philosophy*, 144.
11. Abram, *Becoming Animal*, 108.
12. Ibid.
13. Evernden, *Natural Alien*, 126.
14. Abram, *Becoming Animal*, 109.
15. Evernden, *Natural Alien*, 139.
16. Quoted in Rubenstein, *Strange Wonder*, 28.
17. See Rubenstein, *Strange Wonder*.
18. Rubenstein, *Strange Wonder*, 29.
19. My translation. The original goes: "So ist das Erstaunen die Dis-position, in der und für die das Sein des Seienden sich öffnet. Das Erstaunen ist die Stimmung, innerhalb derer den griechischen Philosophen das Entsprechen zum Sein des Seienden gewährt war." Heidegger, *Was Ist Das—die Philosophie?*, 39–40.
20. My translation. The original goes: "in der Weise des Entsprechens, das sich abstimmt auf die Stimme des Seins des Seienden." Heidegger, *Was Ist Das— die Philosophie?*, 43–44.
21. Abram, *Becoming Animal*, 9.
22. Ibid., 260–261.
23. It is partly due to Earth's gravitational embrace that the infinite space beyond *needs not* invoke a sense of terror. For the attraction between the Earth's larger body and our smaller body is strong enough for us to never have to worry about *falling off*. Earth's commitment to hold us is unyielding.

24. Evernden, *Natural Alien*, 137.

25. Ibid., 141.

26. See Foltz, *Inhabiting Earth*, 158.

27. Quoted in Christie, *The Blue Sapphire of the Mind*, 10.

28. All of the writers I mentioned above are both poets and social critics, many of them activists. What they share is not only their craft but an awareness for the ways in which modern, civilized humans have alienated themselves from that larger chorus. Because of this, their aesthetic project is *also* a moral project: to attempt nothing less than an imaginative reweaving of humans with the more-than-human community, with implications that far exceed the realm of poetry to include the philosophical imagination as much as the historical, the political, the economic, and the legal. To what end? To create a poetics of a planet once again experienced to be sacred.

29. Evernden, *Natural Alien*, 140.

30. Abram, *Becoming Animal*, 63.

31. See Abram, "Reciprocity."

32. She writes further: "The quality of light by which we scrutinize our lives has direct bearing upon the product which we live, and upon the changes which we hope to bring about through those lives. It is within this light that we form those ideas by which we pursue our magic and make it realized. This is poetry as illumination, for it is through poetry that we give name to those ideas which are, until the poem, nameless and formless—about to be birthed, but already felt. That distillation of experience from which true poetry springs births thought as dream births concept, as feeling births idea, as knowledge births (precedes) understanding." Lorde, "Poetry Is Not a Luxury."

Chapter 6

1. In foreword to Margulis and Sagan, *Microcosmos*, 12

2. Ibid.

3. This chapter has been informed substantially by a spiraling series of conversations with David Abram, whose guidance I acknowledge with gratitude.

4. Michaels, *The Winter Vault*, 23.

5. Oliver, "Sleeping in the Forest."

6. Whitman, *Leaves of Grass*.

7. Donne, "LXXX Sermons Preached."

8. Shakespeare, *Family Shakespeare in Ten Volumes*, 19.

9. Montgomery, *Dirt*, 27.

10. Ibid., 4.

11. Ibid., 2.

12. Curry, *Ecological Ethics*, 233.

13. Evernden, *Natural Alien*, 141.
14. *Oxford English Dictionary*.
15. Abram, *Becoming Animal*, 70.
16. Many thanks to Abram for pointing this out. Personal conversation.
17. Abram, *Spell of the Sensuous*, 303n2 (emphasis added).
18. Ibid.
19. Ibid. (emphasis added).
20. Given that the echolocation of bats indeed draws our imagination to its outermost reaches, to a place where perception is diluted more and more strongly by speculation, it is unsurprising that the question of what it would be like to be them has engaged even professional philosophers.
21. Radios, in this sense, extend our capacity for sensual participation with presences that emerge some orders of magnitude more rapidly than that to which we ordinarily are able to tune in.
22. See Matthews, "Letting the World Grow Old."
23. In *The Spell of the Sensuous*, Abram discusses at length the perceptual upheavals brought about by Alphabetic writing. He shows that after the introduction of the new tool that was Alphabetic writing, meaning gradually evaporated from among the flows of the land and condensed increasingly around the inelastic, firm symbols of the alphabet. Abram argues that as a consequence, thought itself seems to have rigidified.
24. See Abram, *Becoming Animal*, 282.
25. There have been some attempts in ecological ethics to bring other animals and plants into the moral circle by considering them "moral patients." While the intention of such an inclusion is plausible, the notion of "moral patients" upholds a certain epistemological bias, namely that we cannot encounter them directly as other-than-human agencies fully able to draw us into a mutually participatory relationship.
26. See Waldau and Patton, *A Communion of Subjects*.
27. See Weston, "Multicentrism: A Manifesto."
28. See Simard, *Mother Tree*.
29. Ibid.
30. Curry, *Ecological Ethics*, 152.
31. Lovelock, *Revenge of Gaia*, 24.
32. Ibid.
33. Abram, *Becoming Animal*, 189.
34. Ibid., 188.
35. Ibid., 189.
36. Brown, "Paving the Planet."
37. This understanding grew out of conversations with David Abram.

Chapter 7

1. Quoted in Grober, *Die Entdeckung der Nachhaltigkeit*, 26 (author's translation).
2. See Grober, *Die Entdeckung der Nachhaltigkeit*.
3. Quoted in Hanh, *Love Letter to the Earth*, 58.
4. Quoted in Grober, *Die Entdeckung der Nachhaltigkeit*, 29 (author's translation).
5. In Hanh, *Love Letter to the Earth*, 23.
6. Quoted in Grober, *Die Entdeckung der Nachhaltigkeit*, 22 (author's translation).
7. Not unlike chapter 6, so, too, this chapter as a whole can be seen as practicing a simple gesture, an attempt to encounter the sea and those who dwell therein ethically. If, as we saw, taking a bow is a gesture appropriate to encountering the land and of placing ourselves in relation to it, then what gesture of thought might entice us into approaching the oceans more ethically? It might simply be this: to cultivate the gift of attention, to bear witness.
8. See Harding and Margulis, "Water Gaia." Unless noted otherwise, the following discussion on water and Gaia theory is based on this article.
9. See Harding and Margulis, "Water Gaia."
10. Ibid.
11. Ibid.
12. The concept of "deep time" was originally coined by the Scottish geologist James Hutton (1726–1797). Most recently, young filmmaker Noah Hutton has released an award-winning documentary named *Deep Time*, a fine attempt to interpret some chapters within geostory artistically.
13. Harding, *Revenge of Gaia*, 185.
14. See Harding and Margulis, "Water Gaia."
15. Harding, *Revenge of Gaia*, 77.
16. Latour, *Facing Gaia*.
17. See Zimov et al., "Permafrost and the Global Carbon Budget"; and Schmitz et al., "Animating the Carbon Cycle."
18. Schmitz et al., "Animating the Carbon Cycle."
19. Latour, *Facing Gaia*, 3.
20. Ibid.
21. Ibid.
22. See Harding, *Animate Earth*.
23. Quoted in Carson, *Sea Around Us*, 213.
24. See Schätzing, *Nachrichten aus Einem Unbekannten Universum*.
25. See Harding, *Animate Earth*.
26. Thomas Berry consolidates the ancient *physis* with contemporary *physics*, in an allusive union of ancient animism and animistic science: "Because creatures in the universe do not come from some place outside it, we can only think of the universe

as a place where qualities that will one day bloom, are for the present hidden as dimensions of emptiness . . . qualities arise from emptiness, from the latent, hidden nothingness of being." (*Story of the Universe*, 76) This resonates with Gaian science: The living Earth has indeed *birthed itself* in an ongoing, several-billion-year-old, gradual emergence out of primordial chaos and into the self-composed sphere it is today. Gaia has composed itself from within the latent, hidden nothingness of primordial chaos, through increasing degrees of integration. It has done so without an outside maker, but solely by virtue of its own emergent complexity.

Such still-recent scientific discoveries resonate with the ancient Greeks' direct, bodily experience of living inside a self-birthing reality. In this, Gaian science has the potential to help reconcile the four-centuries-old split of the mind from the body, sowing some trust again between the rational intellect and the bodily senses This land on which we stand, this soil that shapeshifts to become our food, this air that we breathe, is self-birthed, autopoietic. Nothing is constant. Everything emerges or presences itself, never already there but in a process of ceaseless becoming.

27. See Foltz, *Inhabiting Earth*, 158.
28. Abram, *Becoming Animal*, 84.
29. Quoted in House, *Totem Salmon*, 14.
30. Christie, *Blue Sapphire of the Mind*, 231.
31. Ibid., 226. The quote continues as follows: "Unless we can imagine this possibility, unless we can sense and feel the movement of life across boundaries, and come to know ourselves as participating in this mysterious exchange, we will be condemned to live a thin, impoverished existence, bereft of intimacy, empty of feeling and spirit. And we will almost surely continue to visit our own sense of alienation upon the living world."

Chapter 8

1. Gadamer, *Truth and Method*, 302.
2. See Harding, *Animate Earth*; and Margulis and Sagan, *What Is Life?*
3. Abram, *Becoming Animal*, 293ff.
4. Ibid., 296f.
5. Ibid., 296.
6. Merleau-Ponty was the first to observe this. He said: "As I contemplate the blue of the sky I am not [dualistically] *set over against* it as an acosmic subject . . . I abandon myself to it and plunge into this mystery, it 'thinks itself within me' [i.e., as I lend my senses to it]; I am the sky itself as it is drawn together and unified, and as it begins to exist for itself; my consciousness is saturated with this limitless blue." Quoted in Fisher, *Radical Ecopsychology*, 198n40.

7. Meanwhile we remain fully aware that there will always be some loops with which we are not yet familiar, and also some that we will never learn about, for they remain soundly beyond our perceptual horizon. We also remain aware that some loops of which we think we have a good grasp might one day turn out to be qualitatively very different from what we thought.
8. Personal conversation.
9. Abram has asked the question. In a metamorphic passage on the thinking of birds, he writes: "Flying is an uninterrupted improvisation with an unseen and wildly metamorphic partner . . . It's a kinetic conversation in the uttermost thick of the present moment," he writes, and further that birds are "thinking exquisitely along the whole length of their extended limbs. . . . When we disparage the intelligence of birds, or the size of their brains, we miss that flight itself is a kind of thinking, a gliding within the mind." Abram, *Becoming Animal*, 190–191.
10. Gagliano et al., "Towards Understanding Plant Bioacoustics" (emphasis added).
11. Ibid., 223–224.
12. Ibid., 325 (emphasis added).
13. Ibid., 225.
14. See Leopold, *Round River*.
15. Quoted in Fisher, *Radical Ecopsychology*, 9.
16. House, *Totem Salmon*, 99.

Chapter 9

1. Aldwell, *Conquering the Last Frontier*, 69.
2. Ibid., 20.
3. Ibid., 174.
4. Ibid., 161.
5. Ibid., 182.
6. Ibid., 68.
7. Ibid., 80.
8. Ibid., 77.
9. Quoted in Mapes, *Breaking Ground*, 68.
10. Aldwell, *Conquering the Last Frontier*, 14.
11. Sam Ulmer's telling of the story of how the Klallam tribe came to be called the "Strong People" can be found at "Native Languages of the Americas: Klallam Indian Legends," http://www.native-languages.org/klallam-legends.htm, accessed May 9, 2017.
12. See Valadez, *Elwha River and Its People*. It might not be a coincidence that the Klallam seek guidance for their future in the very place that houses their deepest memory, and their earliest understanding of identity, and of belonging.

Where the experience of temporality cannot be meaningfully separated from the experience of spatiality, where both time and space are aspects of specific places, it is not at all implausible that *any* aspect of time—past, present, and indeed future—might be equally immanent in place.

13. My gratitude goes out to Chuck Larsen for kindly sharing this story with me. Personal conversation.

14. See Mapes, *Breaking Ground;* and Mapes, *Elwha*.

15. I borrow the expressions "original humans" and "new humans" from Maracle, "Where Love Winds Itself," 160–179.

16. See White, *Organic Machine*.

17. As this chapter suggests that the notion of history is closely entangled with the modern form of narrative exceptionalism, I have sought to resist the urge to historicize and to arrange the material about the Klallam strictly chronologically. Instead, I move back and forth between two narrative traditions—that of settler culture and that of the Klallam—using *place* rather than time to make the chapter coherent. In doing so, I continue to practice what Holmes Rolston III has described as the core work of ecophilosophers: to follow tracks through the living world, to pursue narrative careers into the land. Those seeking a chronological history of the settlement of the Elwha specifically, and the Pacific Northwest more generally, might consult Crane, *Finding the River*; Mapes, *Breaking Ground*; Mapes, *Elwha*; Boyd, "Change Is Coming"; Lichatowich, *Salmon without Rivers*; Haeberlin and Gunther, *Indians of Puget Sound*; Colombi and Brooks, *Keystone Nations*; Wolf and Zuckerman, *Salmon Nation*; Wray, *Native Peoples of the Olympic Peninsula*; and Ruby et al., *Guide to the Indian Tribes*.

18. The phenomenon of narrative hegemony certainly did not originate with modernity; neither is it exclusive to modernity. It has deep and knotted roots. Howard Zinn's *Peoples' History of the United States* contains a host of enlightening source material on this theme.

19. Quoted in Allenby and Sarewitz, *Techno-Human Condition*, 18.

20. Abram also discusses this experience of time in "Storytelling and Wonder."

21. See Adamson, *Folk Tales of the Coast Salish*; Clark, *Indian Legends of the Pacific Northwest*; Grinnell, *Grandma's Stories*; Valadez and Charles, *Tse-whit-zen*; Valadez, *Collection of Klallam Stories*; Valadez, *Elwha River*; Valadez, *Boston Charlie and Bigfoot*; Valadez, *Naturally Native*; Valadez, *Klallam Stories*; Smith, *Boston Charlie and Bigfoot*; and Charles, *Klallam Stories*.

22. Mapes, *Breaking Ground*, 21.

23. Name changed by the author.

24. Richard Nelson's descriptions of Koyuokon culture in present-day Alaska describe a people equally embedded in a world that is infused with voices and other-than-human ethical actors. Quoted in Peterson, *Being Human*, 103–105.

25. Indeed, this curious phenomenon has left its track mark in the English language: The northern edge of the Olympic Peninsula is thought of as lying in the mountains' *rainshadow*.
26. See Crane, *Finding the River*, 32.
27. See Gunther, "An Analysis of the First Salmon Ceremony."
28. David Abram first invited me to join reflections around the relationship between space, time, place, and experience in the spring of 2014.
29. Quoted in Deloria, *God Is Red*, 84–85.
30. Ibid., 65–66.
31. Ibid., 87.
32. Ibid., 141.
33. Peterson, *Being Human*, 108. Peterson elaborates: "[Changing Woman] provides the image in which Navajo culture models and constructs persons. This process of construction means that all members of the Navajo nation are related to each other as kin, pointing to the relational definition of what it means to be a 'person' in Navajo culture. Mamie Salt, a Navajo elder, summarizes these relationships: 'The Holy People put us here and the people considered all things sacred, the land, plants, and animals—we are their children'."
34. Not all will be comfortable with my choice of personal pronouns here; surely some would rather see me avoid the personification and settle for the more "neutral" pronoun, *it*. I do not share this concern. Having grown up with German as my mother tongue, I find it quite natural to associate the sun—*die Sonne*—with certain feminine qualities.
35. Capra, *Tao of Physics*, 77–78.
36. It is not unproblematic to use this numbering convention, because the mere mentioning of what is commonly thought of as the year 1592 is already laden with certain assumptions. To say that first contact was "in 1592" already situates the event within the notion of history (and, by inference, not within the articulateness of place-time). While this might be true from one perspective, that of the new humans, it also has a tendency to displace the other perspective, that of the original humans. The conventional use of the Christian calendar tacitly assumes that the perception of time familiar to the new humans is universal, and ought to be universally applied.
37. See Egan, "Restoring the Elwha," 52.
38. See Walbran, *British Columbia Coast Names*.
39. Valadez, *Naturally Native*, 24.
40. What is known today as Hood Canal.
41. See Johnsen, "Salmon, Science, and Reciprocity."
42. See Mapes, *Breaking Ground*.
43. See Crane, *Finding the River*.

44. http://www.northwestplaces.com/ORState/OrState01.htm, accessed October 1, 2012 (emphasis added).

45. Thanks to Doug McDonnell and Chuck Larsen for sharing this.

46. Doug McDonnell and Chuck Larsen also shared this part of the story.

47. Aldwell, *Conquering the Last Frontier*, 120.

48. Quoted in Mapes, *Elwha*, 55.

49. Ibid., 56.

Chapter 10

1. The story of Salmon Boy is told in variations up and down the Pacific Coast. This particular version of it is from the Klallam canon. Thanks to Andrew Fitzgerald for sharing it with me. Personal conversation.

2. See Rust and Totton, *Vital Signs*.

3. Hyde, *The Gift*, 28.

4. Ibid., 29.

5. Deloria, *God Is Red*, 89.

6. Abram, *Becoming Animal*, 254

7. Schiller, "Was Heißt und Zu Welchem Ende Studiert Man Universalgeschichte?" (author's translation).

9. See Masson and McCarthy, *When Elephants Weep*, 27f.

10. Gill et al., "Pleistocene Megafaunal Collapse; see also Barnosky et al., "Assessing the Causes of Late Pleistocene Extinctions."

11. Martin, *Twilight of the Mammoths*, 3.

12. Ibid., 1–2.

13. House, *Totem Salmon*, 99.

14. Personal conversation.

15. In *Finding the River*, Crane delineates the argument, which was first made by the historian Joseph Taylor.

16. See Crane, *Finding the River*.

17. Quoted in Osawa, "The Politics of Taking Fish," 134.

18. Ibid.

19. See Campbell and Butler, "Archaeological Evidence."

20. Ibid.

21. Ibid.

22. Crane, *Finding the River*, 33 (emphasis added).

23. See Catlin, "Letters and Notes."

24. Buhner, *Lost Language of Plants*, 281. The quote goes on: "We moderns, however, in our arrogance and 'enlightenment', have ridiculed such practices, attempted to assign them to the realm of superstition. Ritual has become

'empty ritual'. Thus our connections are in tatters and the world torn asunder. Having ridiculed such rituals, we did not participate in them; not participating in them, we lost our place in the world. And now, how are we to recover our ecological self? Mere ecological ideas, no matter how deep, cannot save us."

25. Berry, *Life Is a Miracle*, 54.

26. See Crane *Finding the River*, 23–24.

27. Jensen, *Endgame*, 186.

28. See Aas, "Våre Fjerne Slektninger i Himalaya," (author's translation).

29. Ibid.

30. Hyde, *The Gift*, 29.

31. Evernden, *Natural Alien*, 135.

32. Likewise, we can also think of all other animals now inhabiting Earth as evolutionary contemporaries to ourselves. Not more or less highly evolved, each of us animals presently living inside the one planet is creatively fitting itself into the morphing texture of the whole.

Chapter 11

1. See PETA, "Commercial Fishing."

2. See SeaWeb, "Bottom Trawling Impacts."

3. See, for example, Amos, "Bottom Trawling."

4. Since the 1950s, fishing fleets have also expanded by focusing increasingly on so-called "nontraditional species." A report by the US General Accounting Office saw the potential to exploit oceanic creatures who had not until then become the target of industrialized fishing fleets, and it identified certain obstacles to the expansion, including "inferior product quality" and "lack of consumer acceptance." But the report had an idea: "Marketing is the key for development of nontraditional fish species." US General Accounting Office, "Developing Markets for Fish."

So it was that the Alaska pollock (*Theragra chalcogramma*), who belongs to the cod family Gadidae and inhabits the coastal waters of the Northern Pacific, came to be marketed by the German fishing industry under the name *Alaska seelachs*, or "Alaska sea salmon," despite having no biological relationship to any of the salmon species. Before fishers started pursuing pollock commercially in 1984, they would throw them overboard as unsolicited by-catch. But within twenty-five years of the first commercial landings, the Alaska pollock had assumed an important role in worldwide fish consumption, with annual catches of about 3.5 million tons by 2010 (about three and a half times the consumption of farmed Norwegian salmon that same year). In Germany, Alaska pollock has gone from being entirely unknown to becoming the number one food fish.

NOTES

Consumers never see "Alaska sea salmon," the fish—the creature with a sentient body, with eyes, a mouth, gills, fins. By the time it reaches their grocery stores, it has undergone a profound shapeshifting of its own. On factory ships far out at sea, the fish are flash-frozen immediately after they are caught, morphing the living creature into uniform fillet blocks. It might be months before these ships eventually return to a port. Once on land, the blocks are transported to processing sites, where they are turned into the frozen consumer end products that reach the markets and restaurants—fish fingers, factory-made meals, or fish patties used by fast-food chains. They might even have been transformed enough to look like salmon. They will be filleted, then boned, then marinated with salt. Then the fillets will be cut into thin slices, dyed with Sunset FCF (E110) and Ponceau 4R (E124) to achieve a salmonlike, red-orange color, then they will be smoked or provided with smoke aroma, and then pickled in brandy vinegar and plant oil.

Another favorite of German consumers, *lachsforelle* or "salmon trout," also has no relationship to real salmon. "Salmon trout," a brochure explains, "is a product name for large, red-fleshed and fatty [farmed] rainbow trout." To turn a rainbow trout into a "salmon trout," the industry adds synthetic dyes to the feed a few weeks before slaughtering the animals. The dyes do not augment the taste of the fish or improve the quality of the flesh. Tests, however, have shown that consumers will more readily purchase the product if associations to salmon can be made. "Salmon" is a catchword in the literal sense. See Reiter, "Die Lachsforelle."

5. Ye and Cochrane, "Global Overview of Marine Fishery Resources," 6.
6. Rockström et al., "Planetary Boundaries: Exploring the Safe Operating Space for Humanity," 32.
7. Feely et al., "Carbon Dioxide and Our Ocean Legacy."
8. See Ettinger, *Sea Change*.
9. Dr. Edward Miles of the University of Washington, in Ettinger, *Sea Change*.
10. See Heinz Awards, "Richard A. Feely."
11. Fabry et al., "Impacts of Ocean Acidification."
12. Ibid.
13. Ibid.
14. Janet L. Armstrong et al., "Distribution, Size, and Interannual, Seasonal and Diel Food Habits of Northern Gulf of Alaska Juvenile Pink Salmon, *Oncorhynchus gorbuscha*," *Deep Sea Research Part II: Topical Studies in Oceanography*, vol. 52 (January 2005): 247–265, https://doi.org/10.1016/j.dsr2.2004.09.019.
15. Richard Feely, quoted in Warren, "Ocean Acidification."
16. Barnosky et al., "Assessing the Causes of Late Pleistocene Extinctions," 56.
17. See Burgess et al., "High-Precision Timeline."

— 325 —

18. Again I think of Gadamer's observation that "[a] person who has no horizon is a man who does not see far enough and hence overvalues what is nearest to him." As I am writing this, I read that Olympic organizers in South Korea—having declared the 2018 Winter Olympics event to be a "Green Olympics"—have clear-cut parts of an ancient Wangsasre forest, situated on a mountain that has been considered sacred to Koreans for centuries. The forest is now being scarred to give way to a large ski resort. The project of modernity, in a nutshell: Five-hundred-year-old-trees are being cut down to make space for a four-day global entertaining event.

19. Quoted in Allenby and Sarewitz, *The Techno-Human Condition*, 18.

20. Quoted in McConnell, "Whittaker's Correlation."

21. See Fromm, *Marx's Concept of Socialism*.

22. Snow, *Two Cultures*, 8.

23. See *Oxford English Dictionary*.

24. Ibid.

25. Ibid.

26. Ibid.

27. Ibid.

28. Ibid.

29. Doney et al., "Ocean Acidification."

30. Ibid.

31. Fabry et al., "Impacts of Ocean Acidification."

32. Ibid.

33. Foltz, *Inhabiting Earth*, 14–15.

34. Berry, *Story of the Universe*, 76.

35. Thoreau, *Walden*, 111.

36. Foltz, *Inhabiting Earth*, 101.

37. A recent debate in the Norwegian media about a hospital reform illustrates this. Hospitals were suddenly being spoken of as "businesses," and patients as "biomass," in a thrust to re-create the Norwegian health sector through a pervasive "logic of industrialization." See Gåsvatn, "Sykehuset med Fabrikklogikk."

38. See Todal, "Spår Eksplosiv Oppdrettsvekst."

39. See Shepard, *Nature and Madness*; Mander, *In the Absence of the Sacred*; and Glendinning, *My Name Is Chellis*.

Chapter 12

1. Kristoffersen, "Vil Satse Bærekraftig," 9.

2. Erna Solberg, NRK P2 radio, March 23, 2015, https://radio.nrk.no/direkte/p2, (author's translation).

3. Latour, *Facing Gaia*, 4.

4. Abram, *The Spell of the Sensuous*, 303n2.

5. Gray, *Heresies*.

6. Ibid., 3–4.

7. Ibid., 4.

8. Capra and Luisi, *Systems View of Life*, 272f.

9. I think of the poetic here, as I have indicated before, in continuation of Heidegger. Foltz writes of Heidegger's thinking on the poetic: "[Poetic] saying is precisely the *logos* of the early Greeks: the gathering together and letting-lie-before us of that which gathers itself and already lies before us, of being as *Logos*. It is possible for us to speak only because we are always already spoken to . . . Poetic saying, then, is 'not just any saying, but that through which everything first steps into the open.'" Foltz, *Inhabiting Earth*, 158.

10. However, I generally hesitate to make any such claim, because I have seen how many times this particular argumentative structure has been exploited to vindicate human dominance.

11. There is one exception. Ecosystems produce one form of waste, which is heat energy of respiration. This energy is radiated into the atmosphere and so is lost to the ecological communities. The loss is counterbalanced by the continuous inflow of more heat energy by the sun. See Capra and Luisi, *Systems View of Life*, 343.

12. Bergljot Børresen has pointed out that this notion of "attentive collaborators," or "co-workers," first emerged in the theology of the Catholic monk Thomas Aquinas (1225–1274). In Aquinas's theology, the term describes people's obligation to collaborate with the Christian creator. Personal conversation.

Chapter 13

1. Evernden, *Natural Alien*, 141.

2. Latour, *Facing Gaia*, 5.

3. Miller, "The Year It Came Apart," 30.

4. Latour, *Facing Gaia*, 3. The name "geostory" would seem tolerable not only for those familiar with the still-emerging Gaian science, but also for those who do not make the direct (etymological) connection between "Gaia" and "geo."

5. Latour says: "Humans are not stupefied to learn that there is an entirely New World at their disposal, but that they have to entirely *relearn* the way in which they inhabit the Old World. This is why in so many ways we feel transported back into the climate of the 16th century. Another age of discovery." Latour, *Facing Gaia*, 6.

6. One who has recently taken bold steps toward reclaiming aspects of that inwardness is Andreas Weber. In his book *Matter and Desire* (2017), Weber speaks of an inwardness that emerges from acknowledging the fact that all beings—perhaps even all matter—experience meaning and significance and respond to these experiences through acts that have to be called "interpretive." Weber suggests that life might simply be impossible without such experiences of meaningfulness.

7. See Latour, *Facing Gaia*.

8. Ibid., 3.

9. Lovelock, *Revenge of Gaia*, 59.

10. Ibid., 59.

11. Latour, *Facing Gaia*, 3.

12. Ibid.

13. Ibid.

14. Ibid., 6.

15. Latour describes such earthbound scientists as follows: "Earthbound scientists are fully incarnated creatures. They are a people. They have enemies. They belong to the soil drawn through their instruments. Their knowledge extends as far as their ability to expand, to finance, to survey, to maintain the sensors that render visible the consequences of their actions. They have no qualms confessing the tragic existential drama in which they are engaged. They dare saying how afraid they are, and in their view such a fright increases rather than diminishes the quality of their science. They appear clearly as a new form of non-national power having a stake in geopolitical conflicts. If their territory knows no national boundary, it's not because they have access to the universal, but because they *keep bringing in new agents* to be part and parcel of the subsistence of other agents. Their authority is fully political since *they represent agents that have no other voices and who intervene in the life of many others*. . . . Freed from the damning obligation of being priests of a divinity they don't believe in, they might even proudly say, 'We are from Gaia.' Not because they entrust themselves to the final wisdom of a super entity, but because, at last, they have abandoned the dream of living under the shadow of a super entity. Secular. Fully secular." Latour, *Facing Gaia*, 5 (emphasis added).

16. Margulis and Sagan, *Microcosmos*, 101.

17. Ibid., 108.

18. Ibid., 109.

19. Chopra and Lineweaver, "The Case for a Gaian Bottleneck."

20. What else could that story be called but *geostory*, which means, simply, Gaia's story? The prefix *ge-*, as in geography or geometry, is an etymological fragment of the ancient, mythological Gaia, preserved into our present time.

Here is how Latour describes what geostory means to him: "The blue planet suddenly stands out as what is made of a long concatenation of historical, local, hazardous, specific and contingent *events* as if it were the temporary outcome of a *'geohistory'* . . . that is, exactly what was not to be taken into account when considered simply as a falling body among all the others." The biosphere has delicately composed itself, as we have already seen, through a "contingent cascade of unforeseen events." Gaia, in this, is Latour's shorthand for "all the intertwined unpredictable consequences of the dispersion of agents, each of them pursuing its own interest by manipulating its own environment for its own comfort—some agents happening to act as a negative and unforeseen feedback on the development of others." Latour, *Facing Gaia*, 4.

21. Latour, *Facing Gaia*, 3.
22. The word "unbelievable" indicates some of the difficulties the thinking mind has—if left on its own—to find orientation inside deep time. It easily becomes disoriented among the necessarily abstract and very large scales.
23. Teilhard de Chardin, *Heart of Matter*, 71.
24. Nelson, *Island Within*, 157–158.
25. Ibid., 51.
26. I borrow this expression from Jeannette C. Armstrong. See Maracle, "Where Love Winds Itself," 166.
27. Abram, *Spell of the Sensuous*, 303n2. See the chapter "Turning Towards the Ground" for a longer discussion.
28. Nelson, *Island Within*, 172.
29. Ibid.
30. Ibid., 51.
31. Ibid., 52.
32. Ibid., 56.
33. Ibid., 54.
34. Ibid., 54, 55.
35. Ibid., 55.
36. Ibid., 52.
37. Ibid.
38. Ibid., 57.
39. See Tucker, "Preparing for a New River."
40. Mapes, "Honoring First Salmon."
41. Ibid.
42. Ibid.
43. Tucker, "Preparing for a New River."
44. Tucker, "On the Elwha."
45. Mapes, "Back to Nature."

46. Ibid.
47. Ibid.
48. Mapes, *Elwha*, 25–27, 10.
49. See Waters et al., "Anthropocene."
50. See "Clayoquot Action."
51. See Dolce, "Pacific Northwest Snowpack."
52. See Geiling, "Why Are Hundreds of Thousands?"
53. Ibid.
54. Ou et al., "Responses of Pink Salmon."
55. Ibid.
56. Quoted in Capra and Luisi, *Systems View of Life*, 452.
57. Ibid.
58. As for ecophilosophy, Arne Næss brought much subtlety and originality to
    sounding out this relational expansion of the (small, egocentric) self into what
    he spoke of as the larger Self. See Næss, *Ecology, Community and Lifestyle*;
    Næss, *Deep Ecology of Wisdom*; and Næss, "Self-Realization."
59. Latour, *Facing Gaia*, 6.
60. Abram, *Spell of the Sensuous*, 303n2.

INDEX

MATTRIX DAYTRON

*M*artin **Lee Mueller**, PhD, received his doctorate in philosophy from the University of Oslo in 2016. Before that, he received his master's degree in culture, environment, and sustainability at the University of Oslo's Center for Development and the Environment (SUM). He has previously helped build teaching centers in rural Mongolia, worked as a kindergarten teacher, been an elementary school librarian, and led a wilderness school in the Norwegian forest. Recently he has also been touring as a storyteller to festivals in the United Kingdom and Scandinavia, with a stage performance inspired by this book. "Salmon Fairytale" weaves together poetry, traditional storytelling, and Sami joik music. He lives in Oslo with his partner and daughter.

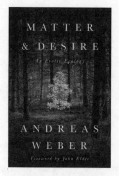

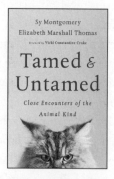
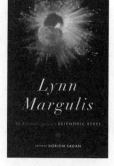